SUSTAINING A NATION

Celebrating 100 years of agriculture in Australia

Jennifer Cornwall • Gordon Collie • Dr Paul Ashton

Focus Publishing
PTY LTD

A FOCUS PUBLISHING BOOK PROJECT

Focus Publishing Pty Ltd
ABN 55 003 600 360
PO Box 518 Edgecliff NSW 2027
Telephone: 61 2 9327 4777
Fax: 61 2 9362 3753
Email: focus@focus.com.au
Website: www.focus.com.au

Chairman: Steven Rich
Publisher and CEO: Jaqui Lane
Associate Publishers: Gillian Fitzgerald
and Jennifer Walkden
Production Manager: Timothy Ho
Project Manager: Gary Deigan
Client Services: Sophie Beaumont

Editor: Philippa Sandall
Designer: Sarah Cory
Consultant: Kathy Walter

© Commonwealth of Australia 2000
This book is copyright. Apart from any
fair dealing for the purpose of private study,
research, criticism or review, as permitted
under the Copyright Act, no part may
be reproduced by any process without
written permission. Enquiries should be
addressed to the publisher.

ISBN 1 875359 74 5

Whilst all reasonable attempts at factual
accuracy have been made, Focus Publishing
accepts no responsibility for any errors
contained in this book.

ABOUT THE PUBLISHER

Focus Publishing Pty Ltd, Australia's leading corporate publisher, specialises in producing high-quality custom and brand books, corporate histories and specific marketing, event, promotional and anniversary books. Focus creates high-quality publications which are used by leading enterprises for strategic marketing, branding and customer loyalty programs.

In addition, Focus provides a range of archiving, oral history and knowledge management services dedicated to ensuring that Australia's corporate history and knowledge are integrated into current corporate communication.

Focus also has an event management division that specialises in creating and managing major corporate events.

Clients include Ansett Australia, Australia Post, Australian Stock Exchange, Commonwealth Bank of Australia, Ford Australia, MBF, Qantas Airways, Snowy Mountains Hydro-electric Authority, State Library of New South Wales, Sydney Airports Corporation, The University of Sydney, Traveland, Westpac and Woolworths Limited.

Visit the Focus website: www.focus.com.au

CONTENTS

Foreword   4

Introduction   5

PART ONE — The Wealth of the Land   8

Meat	10	Horticulture	60
Dairying	22	Fisheries	70
Broadacre Crops	32	Forestry	80
Sugar	42	Fibres	90
Wine	50		

PART TWO — Challenges and Opportunities   100

Life on the Land   102

Getting Goods to Market   114

Sustainable Agriculture   124

PART THREE — Showcase of Participants   134

Roll of Honour   135

www.focus.com.au   190

Index   191

FOREWORD

IT IS WITH GREAT PLEASURE that I endorse *Sustaining a Nation*. This book recognises the importance of agriculture in the growth, development and economic strength of Australia over the past 100 years. It is also a tribute to the men and women of rural Australia and celebrates their dedication, resilience and contribution to this country. Key agricultural organisations and associations have also played an important role in the sector and continue to do so to this day.

Australia is a country defined by its agricultural sector. Agricultural products were among the first goods traded by this country and remain a critical element of our current and future international trade. Our quality of life is enhanced by the wealth generated by the agricultural sector and the clean, green quality of our food and agricultural products.

Australia can rightly claim world leadership in many areas of agriculture in terms of developing better breeds of animals, plants, fish and grains; in its support and commitment to trade liberalisation worldwide and to improving land management and sustainable agricultural, fisheries and forestry techniques and policies. Australia leads the world in water resource management, an area which will be of vital importance in the 21st century if the world is to be able to feed an ever growing population.

The Centenary of Federation celebrations in 2001 are an appropriate time for all Australians to reflect on the vital part that agriculture has had in our lives, and to recognise the importance that a vibrant agricultural sector will have on our continued health, wealth and well being as individuals and as a nation.

Sustaining a Nation is a timely celebration and recognition of Australia's agricultural sector and all those who play a role in it, no matter how small or large.

I commend this book to all Australians.

Hon. Warren Truss MP
Minister for Agriculture, Fisheries and Forestry

INTRODUCTION

'Dear Bert, Just a few lines to you before Christmas. This is a snap of my 22 acres, Gresby, and this is not the highest place … the damn frost the night of the Melrose Show put the kybosh on our wheat again … it just about breaks one's heart to try and grow wheat here. We are in the 100 acre pad next to Bill's, will go about 8 bags … Oats went 13 bus, barley 6 bags. We haven't any carted yet — had to make a set of false combs — made them in a day, the other set I got in 1920.'

— Letter to Bert, 2 December 1932 from unknown soldier settler, New South Wales

THE FIRST WHITE SETTLERS arrived in Australia with high hopes of recreating a rural landscape just like the one they had left behind. Their confidence derived in part from Joseph Banks who, having seen the densely forested eastern coastline, confidently advised the British authorities that the whole continent could produce wheat, fruit, flax, tobacco and cotton with 'tropical abandon'.

But beyond the seaboard, Australia is one of the world's driest continents and farming has always been conducted under difficult conditions. As a result, we eventually became pioneers of dryland farming, exporting our 'know-how' to other parts of the world with similar environmental constraints.

Rural life throughout the 20th century had an enormous impact on the way Australians defined themselves and on the economic success of the nation. At the outbreak of the Second World War, almost a quarter of gross domestic product came from rural industries. Today it represents 4 per cent. But the land and those who work it endure as a sustaining feature of the nation's past and present and the sector continues to be an important source of the nation's export income.

The Centenary of Federation is an appropriate time to look back at the agricultural policies and practices that sustained a nation for a hundred years and to look ahead to sustainable agriculture along with advances in technology and productivity.

As Australia became a nation on 1 January 1901 its future was perceived to be vested in the wealth of its land. Closer settlement schemes — which sought to establish a class of farmers on small land holdings in the tradition of the British yeoman — were to be the principal means of achieving this. These schemes also proved to be among the most significant factors in shaping agricultural development in the 20th century.

ABOVE: A casual worker at Montrose in New South Wales, 1930.

SUSTAINING A NATION

A Handful of Australia

Enthusiastically proclaiming the healthy advantages of rural life, the new state governments pursued the schemes with the same enthusiasm as their colonial predecessors had. Federation, however, added another dimension — nation building. Selectors were cast as the 'dynamos of national economic growth' who would untap the vast wealth lying dormant in the land.

But the closer settlement schemes fell far short of romantic images of rural life and ready markets. Many selectors with very little capital and even less experience struggled to carve out a precarious existence on insufficiently sized holdings often unsuited for the intended agricultural production.

Despite mounting evidence of their failure, governments embarked on rural soldier settlement schemes as a repatriation measure after the First World War and over 37,000 soldiers were repatriated in this way.

'Australia Unlimited', the optimistic catchcry of the times, embodied the idea of an unlimited resource — land — which needed only capital and labor to become productive. The Depression was to expose the fallacy of this as many of the soldier settlers who had managed to struggle through the 1920s, undercapitalised and burdened with debt, faced ruin.

Agricultural expansionist policies had clearly been guided by ideology rather than sound economic reasoning and planning and an understanding of the fragility of the land that they were to plough and farm.

By encouraging mechanisation (where it was available), the Second World War marked a transition from pre-industrialised to capital intensive agriculture and provided state and federal governments with an opportunity to look at various aspects of farming and rural life and re-examine agricultural policies. The Rural Reconstruction Commission established in 1943 recommended larger, capital intensive farms as the key to survival in the postwar world, but as yet there was still no real appreciation of the environmental impact of improved agricultural practices.

In a related and timely development, ABARE's predecessor, the Bureau of Agricultural Economics (BAE), was formed. In the prewar years, there was no research agency to advise government on economic policy relating to the sector. Such advice tended to be obtained on a one-off basis, often through royal commissions. The BAE was established in 1945 and merged with the Bureau of Resource Economics in 1987 to become the Australian Bureau of Agricultural and Resource Economics (ABARE).

One of BAE's first tasks was under the War Service Land Settlement Scheme. Soldier settlement on the land had been abandoned by 1929. A similar scheme revived after the Second World War was implemented with a greater degree of success. The BAE provided advice on the income potential of thousands of settlement proposals, principally through assessing market outlook for the industries involved. These outlook assessments were to form the basis for what has probably become the Bureau's largest and most high profile single activity — the ongoing program of publishing commodity forecasts and projections.

The second half of the 20th century was to be a period of further changing fortunes for the sector. The impact of a more global market was already being felt by the 1960s as advances in agricultural efficiency and protectionism led to excess supply and depressed prices.

Competitiveness in a world market required greater efficiency which in turn spelled increased capital, larger landholdings, diversification and deregulation as many of the bounties and subsidies that had characterised the sector for so long were removed. The small family farm, the key productive unit in agriculture for the first half of the 20th century, was no longer the mainstay of farming as farmers either 'got big' or 'got out'.

While ABARE's core role of collecting, analysing and disseminating information has not changed fundamentally since it was established, it too has had to respond to an ever changing environment, both in terms of domestic developments and environmental research and to changes in international market conditions.

As the initiator and lead supporter of this pictorial history — *Sustaining a Nation* — we have an opportunity to reflect on the people and production processes that have shaped modern agriculture in Australia and to look ahead to finding ways to sustain the nation with sustainable agriculture.

Dr Brian Fisher, *Executive Director*, ABARE

PART ONE

THE WEALTH OF THE LAND

'And now, this first day of January 1901, we enter upon our new era as a budding nation with becoming modesty, to blossom out with an appropriate blush and a pleasant smile — not as the poor miserable little settlement of 112 years ago, half-starved, surrounded by the unknown, half afraid to creep beyond coo-ee of the first-selected spot, but radiant with success, with the strength of a young giant concentrated in our soils, in our climates, and in our bright clear sunshine.'

— WH Clarke, *Agricultural Gazette of New South Wales*, January 1901

The Wealth of the Land

MEAT

'Of course meat is the staple of Australian life ... A working-man whose whole family did not eat meat three times a day would indeed be a phenomenon. High and low, rich and poor, all eat meat to an incredible extent, even in the hottest weather.'

— Richard Twopeny, *Town Life in Australia*, 1883

SELF APPOINTED ADVOCATE of healthier living, Dr Phillip Muskett, declared in 1893 that the consumption of 'butcher's meat' in Australia was 'enormously in excess of any common sense requirements'.

In fact, medical opinion of the day attributed a whole range of maladies to our fondness of meat — heart affections, constipation, liver disorders, hallucinations and even fits of passion. But all this went unheeded by the Australian working man, who enjoyed his meat three times a day — for breakfast, lunch and 'tea'.

By the 1940s, meat was being credited with a 'special dietetic quality' which gave those who ate it 'that more explosive initiative energy'. It wasn't until the 1970s that meat consumption began to drop as eating patterns changed.

The advent of refrigeration in 1880 had opened up a significant trade in beef and mutton. By 1895, Australia was supplying 20 per cent of Britain's meat imports. Mutton had always been cheap and plentiful — it was essentially a byproduct of the more lucrative wool industry. Beef production, on the other hand, had been developing as a specialised market since the gold rushes of the 1850s.

By 1934, refrigerated shipment techniques had advanced and meat producers could now chill their carcasses, allowing them to compete in the overseas market with South America. And with the Ottawa agreements of 1932, under which Australian primary produce had preferential access to British markets, confidence in the beef industry encouraged further expansion.

OPPOSITE: Poll hereford cattle, Dunkeld, Victoria.

This chapter is proudly brought to you by Consolidated Meat Group Pty Ltd *and* Consolidated Pastoral Company Pty Ltd.

SUSTAINING A NATION

ABOVE: 'Dipping' cattle into an arsenic solution to kill cattle tick at Anthony's Lagoon in 1914, possibly in Queensland. Inspection points ('tick gates') were introduced toward the end of the 19th century to stop its spread south. Arsenic solutions were replaced with sprays of highly toxic chemicals by the late 1930s. But these were equally as hazardous leaving a residue in the meat. The cessation of spraying coincided with the introduction of tick resistant cattle breeds such as the Brahman in tropical Australia.

With the negotiation of a meat agreement with Britain in 1952, the beef cattle industry expanded further into the tropical north, which was emerging as the dominant beef producing region.

Refrigeration technology, combined with advances in agricultural science, paved the way for another pastoral industry: prime lamb production. By the 1930s, agricultural scientists had made headway and farmers in the south eastern states reaped the benefits of cross-breeding and improved pastures. Over the next thirty years more than half the prime lamb produced was exported, mostly to Britain, which remained the major market until 1973.

On the home front, Australian tastes in meat were changing. Lamb had replaced mutton as the preferred sheep meat by the 1960s, and commercial poultry production made its debut. The move into mass production resulted in a rapid fall in the price of chicken. Consumption rose from 5 kilograms per person in 1965 to 8 kilograms in 1968 and 31 kilograms in 1999. Chicken meat was now an affordable alternative to red meat, and a popular takeaway food.

By the mid-1960s the landscape changed again for meat producers. For beef producers, the expiry of the British trade agreement coincided with the entry of the United States into the market, looking for lean beef. Lamb and mutton also found new markets in Japan, the USSR and the Middle East. The development of the Middle East as a major market for lamb, mutton and live sheep for slaughter has been a major influence in the industry.

The most dramatic challenges for the red meat industry in the domestic market had their beginnings in the 1970s. Meat consumption had begun to drop — a casualty of dietary concerns, changing eating patterns and a growing interest in vegetarian diets. The industry responded with trim cuts and advertising campaigns, one of which exhorted the homemaker to 'Feed the Man Meat'. White meat producers have had to weather changing consumer tastes too. Pigs — no longer a sideline of dairy farming — are leaner and meaner.

Although Australians are unlikely to return to the meat consumption levels of a century ago, we remain among the biggest meat eaters in the world. And the enduring popularity of that most

important social ritual, the barbeque, secures the place of meat in our way of life.

A larger range of meats than ever before is on offer, including leaner meats, trim cuts and a wide variety of marinated or semiprepared products to keep pace with a faster lifestyle, health concerns and cosmopolitan tastes.

Boutique specialty markets have also emerged for the more exotic products such as emu, kangaroo and crocodile.

And while the local butcher's shop is increasingly superseded by the supermarket, the butcher remains a powerful icon in the industry's arsenal.

ABOVE: The prime lamb producing area extends down eastern Australia from Queensland's Darling Downs, through New South Wales, Victoria and Tasmania, to the Eyre Penninsula of South Australia.

SUSTAINING A NATION

ABOVE: 'Free range' poultry farming in the 1920s. Farmed as a sideline to egg production, poultry remained an expensive delicacy that only appeared in most households on special occasions until the advent of commercial poultry farming and mass production in the late 1950s.

RIGHT: Fooks' Cambridge Hatchery, an early commercial hatchery delivering 'quality day old chicks' to a poultry farm around the late 1930s. The advent of agribusiness technology meant chicken production was moving into the mass production era — the formerly independent poultry farmer now received day old chicks, feed and veterinary supplies from the processor and delivered the grown birds to the same company. Further expansion occured in the 1960s with the rise of vertically integrated companies. Consumption increased with falling prices.

SUSTAINING A NATION

ABOVE: Rabbits being collected from the traps in the 1920s ready for delivery direct to homes or to the factory. While a menace on the land, rabbit was an important part of the Australian diet and a source of income by the 1880s. It also provided the basis for substantial commercial industries. During the First World War rabbit skins were used for the manufacture of military felt hats. The 'rabbitoh' was a common sight in the streets until the 1930s.

ABOVE: Pig farming in the 1930s. Pigs became a common sideline to dairying after the introduction of cream separators in the 1880s as they were fed the skim milk remaining after the separation process. Some were also fed on whey collected from the local cheese factories. Intensive farming and health concerns were among some of the reasons that led to a shift to grain feeding.

RIGHT: Pigs feeding on grain. Pig raising has become a specialised production activity in Australia. With precise carcass specifications relating to fat content demanded by the market, production is increasingly concentrated in the hands of specialist producers who control fat content through breeding and formulation of feed. Before the Second World War a mere 7–8 per cent of total meat eaten by Australians was pig meat. By 1998 this had grown to 20 per cent.

SUSTAINING A NATION

ABOVE: A display of mustering skills. From the 1970s, helicopters have been used to assist in the mustering of cattle in inaccessible areas, especially in the Northern Territory.

RIGHT: From the 1870s, the sparsely vegetated savannah areas of the tropical far north — the Gulf Country of Queensland, the Northern Territory and the east Kimberley district of Western Australia — became cattle raising country. The large holdings were established on the open range system. In more than two-thirds of the area the cattle industry depends for water largely on subartesian and artesian bores. Steady expansion of the industry in the 1960s was driven mainly by the growth of the United States export market. In northern Australia it was a period of optimistic land clearing (such as the Brigalow lands of Queensland) which opened up millions of hectares of good quality pasture.

SUSTAINING A NATION

LEFT: Branding cattle to prevent cattle duffing. While a practice largely associated with the 19th century, cattle stealing continues to this day, with stock squads still maintained in Queensland and New South Wales. The incidence of cattle duffing tends to rise and fall with the price of cattle.

OPPOSITE: Beef cattle grazing in Cootamundra, New South Wales. While Queensland is Australia's largest beef producer, beef cattle production is more intensive in south eastern Australia because of the greater proportion of land under improved pasture. Beef raising in southern Australia is typically combined with cropping and sheep raising. By the 1970s, cattle in northern Australia were raised mainly for export, while in the southern states the beef cattle industry was oriented toward supplying both domestic and overseas markets.

LEFT: A cattle feedlot at Millerman, Queensland. Linked to the demands of overseas markets such as Japan and Korea, the growth of the feedlotting industry in the 1990s has significantly boosted beef and grain production.

20 — SUSTAINING A NATION

SUSTAINING A NATION

The Wealth of the Land

DAIRYING

'... the period 1880–1920 can be characterised as one in which the dairy industry was firmly established as a commercial industry, due to the exploitation of a growing market ... which was in turn assisted by technical developments, economic re-organisation, and establishment of the factory system ...'
— N Drane and H Edwards, *The Australian Dairy Industry: An Economic Study,* 1961

IF THERE IS ONE AREA OF FARMING that epitomised Steele Rudd's Dad and Dave battler on the land, carving out an existence on a small holding through their own labors, it is the dairy industry.

Refrigeration, machinery and the establishment of the cooperative factory system transformed dairy production in the late 19th century. Milking machines appeared in the early 1900s (although they did not become commonplace until the 1950s), significantly reducing the labor intensive nature of dairying. However, many dairy farmers had little capital, and were unable to adopt new technologies. Pasteurisation, also available from the beginning of the 20th century, improved the quality of butter and the hygiene in liquid milk.

Dairy farming expanded strongly up to the First World War under closer settlement schemes. New districts continued to be opened up in the 1920s as soldiers were repatriated on the land under similar schemes. But the optimism was short lived. When the wartime trade agreement with Britain ended in 1921, prices tumbled. There were too many suppliers in the market, and they had been heavily dependent on exports. The high prices for butter in 1919–20 had greatly inflated land values, compounding the problems of overcapitalisation which newer dairy farms were to experience throughout the 1920s.

Price fluctuation was the impetus for political action. The federal government introduced an equalisation scheme known as the Paterson Stabilisation Scheme in 1926. Under this arrangement, the domestic price of butter and cheese rose by the amount

OPPOSITE: Bringing in the milk. Since the 1950s, the Friesian has been favored by most dairy farmers in Australia producing for the liquid milk market.

This chapter is proudly brought to you by The Dairy Farmers Group.

SUSTAINING A NATION

RIGHT: Before the 1930s there was little regulation of the treatment, storage and transport of milk. Milk was a major carrier of disease, such as tuberculosis and brucellosis. Improvements in hygiene and purity with the advent of bottling and sterilisation meant that advertisements like this one from 1937 could confidently advocate the drinking of milk.

OPPOSITE: Blackfriars Infant School, Sydney, 1923. The school milk scheme was instituted to address children's dietary deficiencies. The dairy lobby also played a major role in its establishment. By 1950 a universal scheme was put in place by the federal government. Such was the importance of this market to the nation's dairy farmers that, in Sydney alone, total milk sales were reduced by 2.7 per cent when it was discontinued in 1974.

needed to cover the gap between the return on exports and what was deemed a reasonable return to the producer.

So dairy farming entered a new phase of expansion in the early 1930s. But while price support provided some financial security, it also encouraged inefficient and marginal farmers.

Dairy farmers producing liquid milk fared much better than those producing milk for manufacturing up to the 1990s, when the issue of free interstate trade (under section 92 of the Constitution) and supply monopolies surfaced. The end result was the deregulation of the fresh milk market in 2000. Until then state governments controlled the quality and supply of liquid milk through statutory authorities. Liquid milk was received only from certain designated producing districts, quotas were allocated to farmers, premium prices were paid and retail prices were fixed. And with growing urban populations, the demand for fresh milk was increasing.

Labor shortages during and after the Second World War led to further mechanisation and a period of postwar prosperity. But by the early 1950s the old problems were back: too many farms, insufficient capital to adopt new technologies, a small export market and fluctuating prices.

Consumers began moving from butter to margarine in the 1950s. Government restrictions on margarine production stalled the slide, but these restrictions were abandoned twenty years later. At its peak in 1953–54, annual butter consumption was just under 13.9 kilograms a person: over the next four decades it dropped to 2.6 kilograms.

Prices for dairy produce continued to fall in the 1960s, while production costs, such as fertilisers and machinery, increased. In the 1970s the price support schemes for butter and margarine quotas were removed and assistance was slowly withdrawn, removing the safety net for the small family farm.

Adding to its problems, the industry lost its most important export market in 1973 when Britain entered the European Economic Community.

The good news was the increasing demand for cheese. By the 1960s, cheese was becoming a

RIGHT: A cheese factory around 1910. Cheese was at this time a rather 'soapy' cheddar that was sold as either 'mild' or 'tasty'. In 1926 a Melbourne manufacturer, Fred Walker, secured the Australian rights for the Kraft processed cheese which had excellent keeping qualities. After Walker's death in 1935, the American giant Kraft Foods gained control of his company and went on to dominate the production of processed cheese in Australia.

OPPOSITE: Churning butter from cream in the 1920s. The introduction of cream separators on farms from the 1880s led to the establishment of cooperative butter factories. Butter production was the foundation of the Australian dairy industry. In the late 1920s, out of 821 million gallons of milk produced in Australia, just over three-quarters was turned into butter.

popular part of the Australian diet. And 'fancy' cheeses including gouda, ricotta, cottage, fetta and mozarella were being manufactured locally. Since this time, the consumer market for cheese has continued to diversify and the industry now produces a wide variety of cheeses including mould ripened cheeses such as blue vein, camembert and brie.

In the past thirty years the dairy industry has developed new products for new markets: low fat and iron and calcium enriched milk and yoghurts for the health concerned consumer; vanilla custard and full cream flavored milk for those wanting an indulgent taste sensation. There are even low cholesterol and reduced fat cheeses. By meeting head on the challenges of a market obsessed by the fat content of foods — the natural component of dairy produce — the industry has demonstrated its entrepreneurial skill and amazing resilience.

26 — SUSTAINING A NATION

SUSTAINING A NATION

ABOVE: Dairy farming, possibly in Gippsland, Victoria, around 1910. Victoria is Australia's leading dairy producer. As with the wheat industry, the small family farm was established as the productive unit.

LEFT: Digging a silage pit in the 1930s. Meadow pastures, the primary feed source of dairy cows were harvested and ensiled in large pits like this one, providing a cheap source of feed for the winter months. Unlike meadow hay, the harvest could be thrown straight into the pit without the need to dry it first. The downside was the smell that would emanate from the pit and the mould problems.

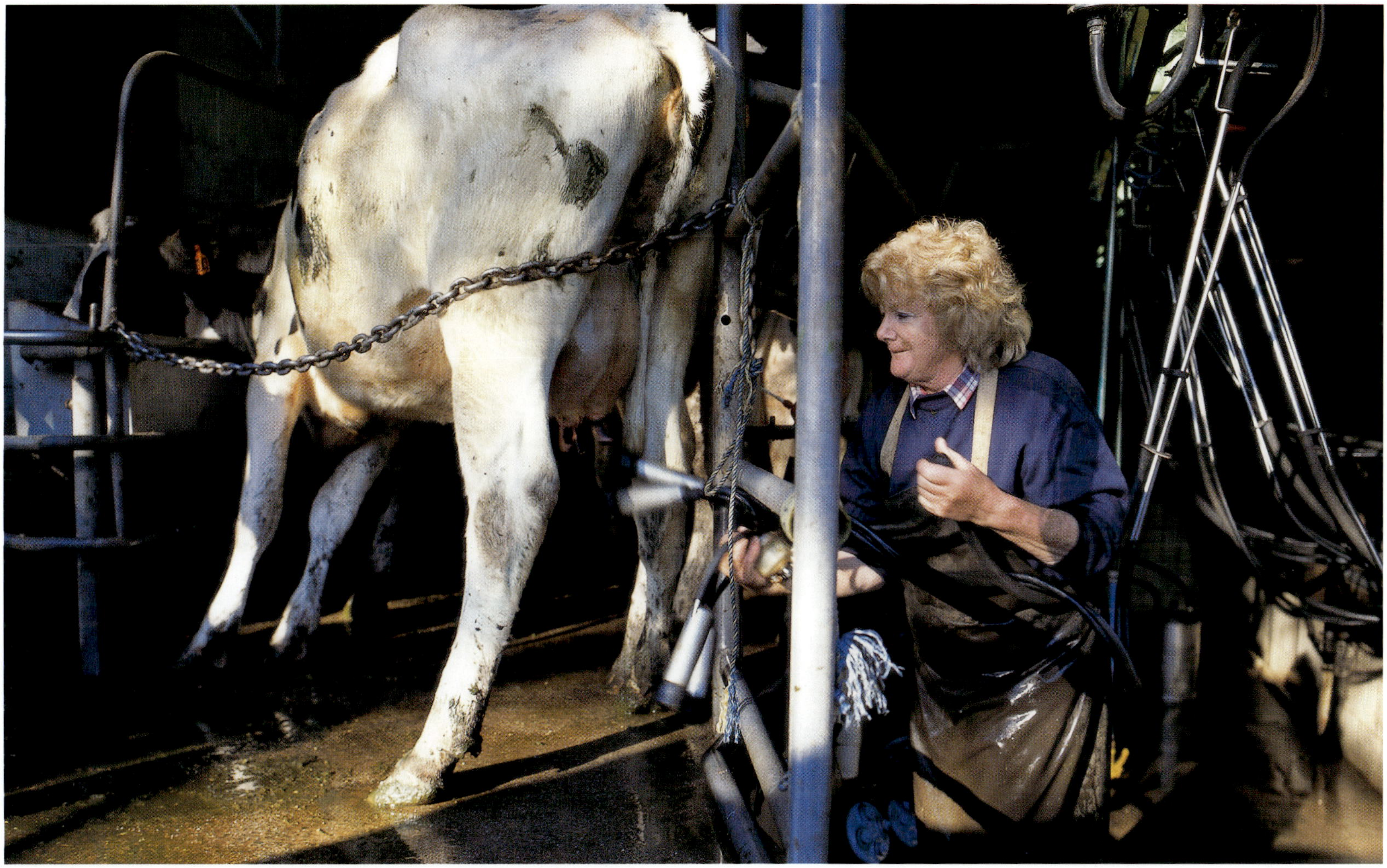

ABOVE: These days, the walk through dairy has been replaced with herringbone and rotary dairies and the use of automatic teat-cup removers.

LEFT: An early milking machine around the 1910s. While relieving much of the drudgery of dairying, it was not until the 1950s that the milking of commercial herds throughout Australia was mechanised. Dairying was labor intensive and often involved all family members. While child labor was a contentious issue in farming generally, it was identified principally with the dairy industry.

SUSTAINING A NATION

ABOVE: 'Hygienic glass lined' milk tanks for railway transport owned by Mr J McFarland outside Dairy Farmers Co-operative Milk Co. Limited in Sydney in the 1930s. The spread of railways to major dairying areas provided fast transport to markets. These milk tanks represented the latest technology and were part of attempts to improve the hygiene and purity of milk in this decade.

ABOVE: The refrigerated tanker has meant that the 10 gallon milk cans left at the farm gates for collection have become a thing of the past.

SUSTAINING A NATION

The Wealth of the Land

BROADACRE CROPS

'Farming was not the drag, the wretched, murderous drudgery, it used to be. We were improving every day, climbing rapidly to the lap of comfort. The wheat turned out a success again, and the profit made us all rejoice. Still we kept our heads …'

— Steele Rudd, *Our New Selection*, 1899

WHEN THE AUSTRALIAN COMMONWEALTH was proclaimed on 1 January 1901 in Sydney's Centennial Park, white flour, sugar, rice, oats and sago would all be found in substantial quantities in the kitchen cupboards of the new nation. Flour, in particular, played the leading role in home baking, reflected in the size of its cannister standing tall at the head of the set. Today home baking is very much a thing of the past, and our main broadacre crops — wheat, barley, oats, sorghum, maize and rice — and the way we consume them illustrate major shifts in the way we live and what we eat.

WHEAT From 1901 until the outbreak of the First World War, the wheat belt almost doubled, with the expansion of rail, roads and closer settlement schemes, particularly in New South Wales and Western Australia. Although many of the new wheat districts proved marginal, wheat yields increased markedly with the introduction of new varieties (William Farrer's high yielding, drought resistant 'Federation' being the best known), mechanisation, the use of superphosphate and fallowing.

Between 1915 and 1922, the government 'Wheat Pool' compulsorily acquired and stockpiled all wheat to meet wartime demand and stabilise prices. The success of this scheme laid the foundation for later government interventions. But the 'Grow More Wheat' campaign in 1930 — an attempt to improve the balance of payments crisis — led to a substantial increase in production, coinciding with a collapse in prices and a regular diet of boiled wheat and treacle ('cocky's joy') on the farmer's table.

Relief for the debt burdened farmers with initiatives such as the 1931 bounty and the 1933 flour tax proved to be merely stopgap measures.

OPPOSITE: Wheat has always been the dominant crop grown in Australia. Wheat areas have grown from 2 million hectares at the time of Federation to over 10 million hectares today.

This chapter is proudly brought to you by Goodman Fielder, NSW Grains Board *and* Ricegrowers' Co-operative Limited.

SUSTAINING A NATION

RIGHT: Barley at Tooths Brewery in 1948.

OPPOSITE: Self contained harvesters in the 1920s, a far cry from the autoheaders with airconditioned cabins in the late 20th century. Manufactured by Hugh Victor McKay's famous Harvester Company from 1887, the Sunshine Harvester combined stripping, threshing and cleansing grain. This meant that harvesting could be completed in a single process by placing a winnower within a stripper, a revolutionary development greatly reducing the labor involved.

Real relief only came with the introduction of the guaranteed home consumption price in 1938, a measure that endured for decades. At the outbreak of the Second World War, a wheat board (the Australian Wheat Board or AWB) was set up to stabilise prices and meet wartime demand. It became permanent in 1948; and its monopoly of the domestic market was to last for forty years.

The security of a minimum price, mixed farming, pasture and soil improvement, more disease resistant varieties and improved cultivation techniques led to a further expansion of the wheat belt. Increased mechanisation — tractors, motorised headers and bulk grain handling systems — increased productivity.

In 1989, domestic wheat marketing was deregulated, giving growers the choice of trading privately or delivering to an AWB pool. Today, the AWB retains responsibility for export, building its marketing strategy around the quality of Australian wheat — white, dry, clean and insecticide free.

RICE The first successful rice crop in Australia was grown in 1914 by a Japanese family near Swan Hill in Victoria. The interest this aroused led to the Yanco Experiment Farm conducting rice growing experiments a year later. Commercial production, made possible by the irrigation of the Murrumbidgee, began in 1924 near Leeton and Yenda. Despite attempts to grow rice in Queensland and Western Australia, the Riverina remains Australia's principal rice growing region.

34   SUSTAINING A NATION

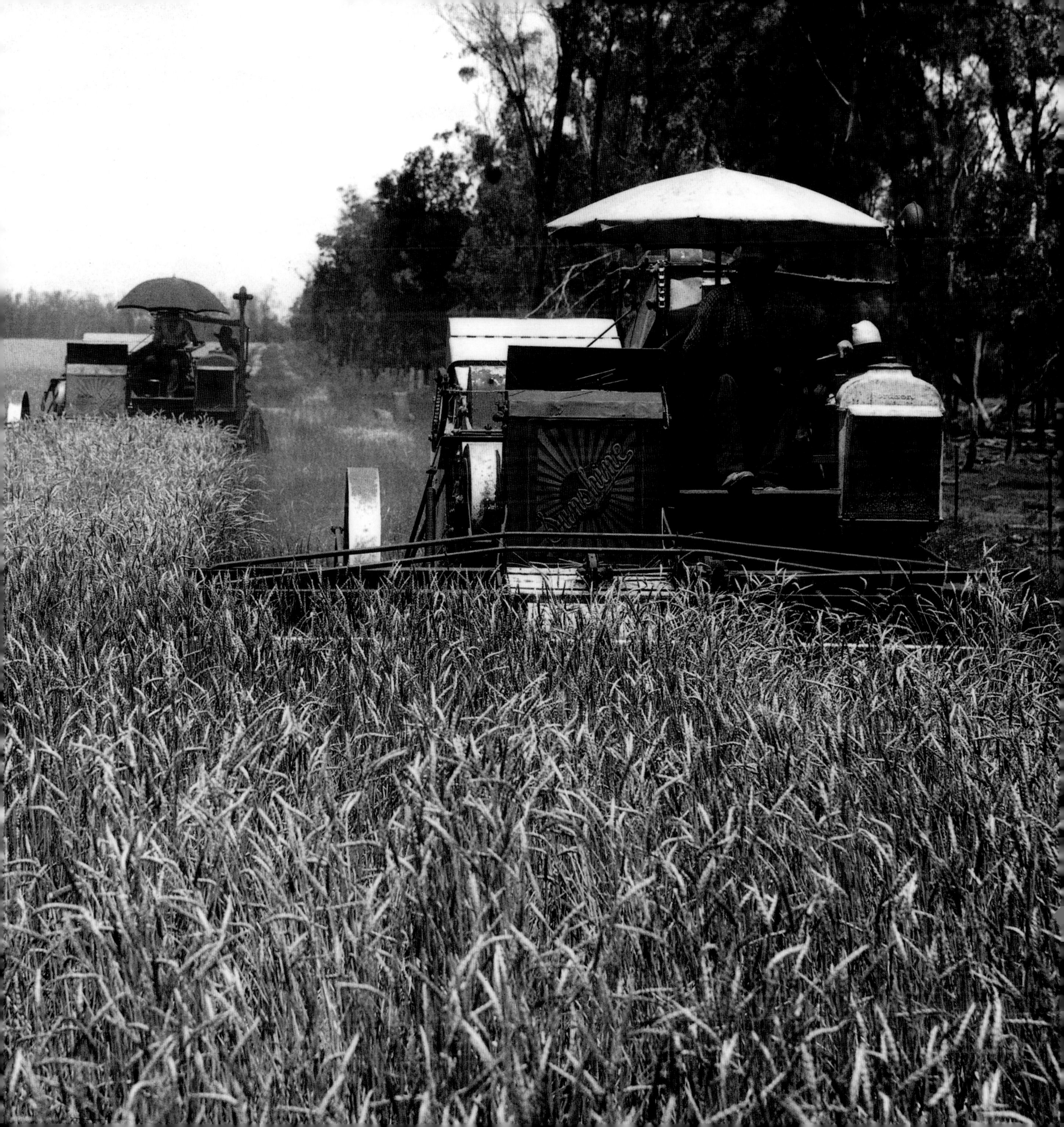
Sunshine

RIGHT: Building haystacks was a specialised skill that has now disappeared. The local 'stackie' moved from farm to farm during the harvest plying his trade. The haystack was constructed at an angled pitch, making it broader at the top than at the base, to keep the sides protected from rain. The last sheaves were used to finish off the stack with a thatched roof effect for waterproofing. From the 1940s, building haystacks was progressively replaced by baled hay and the hayshed.

OPPOSITE: Cutting sheaf hay stored under cover into chaff which is being bagged and sewn to be sold as feed. Note the 'power' drive from the fly on this early tractor. The appearance of the hayshed was to replace the haystack as the means of preserving the hay.

BARLEY The dominant barley growing areas in Australia are the Yorke Penninsula in South Australia and the Southern Riverina in New South Wales. A little under half of Australia's barley production is used for malting and brewing purposes. The remainder is used for feed.

SORGHUM, TRITICALE, OATS AND MAIZE The importance of these grain crops, grown primarily as fodder, increased with the advent of intensive feedlot enterprises (cattle) and grain feeding (pigs, dairy and poultry). Most grain crops have traditionally been controlled by marketing boards, but the focus is shifting to highly differentiated grain and grain products for specific markets including stock feed, bread manufacturing, the health food industry and processed breakfast cereals.

PULSE CROPS With their natural ability to improve soil fertility, these crops have played an important part in sustainable farming since the early days of pasture improvement in the 1920s. Today they are grown for enhancing wheat yields (faba beans), stock feed (lupin and field peas) and human consumption (chickpeas and lentils).

OILSEEDS Although imported linseed was first crushed in New South Wales in 1908, commercial oilseeds growing didn't get underway until 1946 (linseed) and 1955 (safflower). In the 1960s, the industry widened, with the crushing of cottonseed (1963), rapeseed (1966) and sunflower (1968). Today's much larger oilseeds industry crushes canola, cottonseed, sunflower, soybeans, safflower, linseed, linola, peanuts, macadamia nuts, olives and mustard seeds, and includes 'boutique' country based mills processing specialised niche market crops.

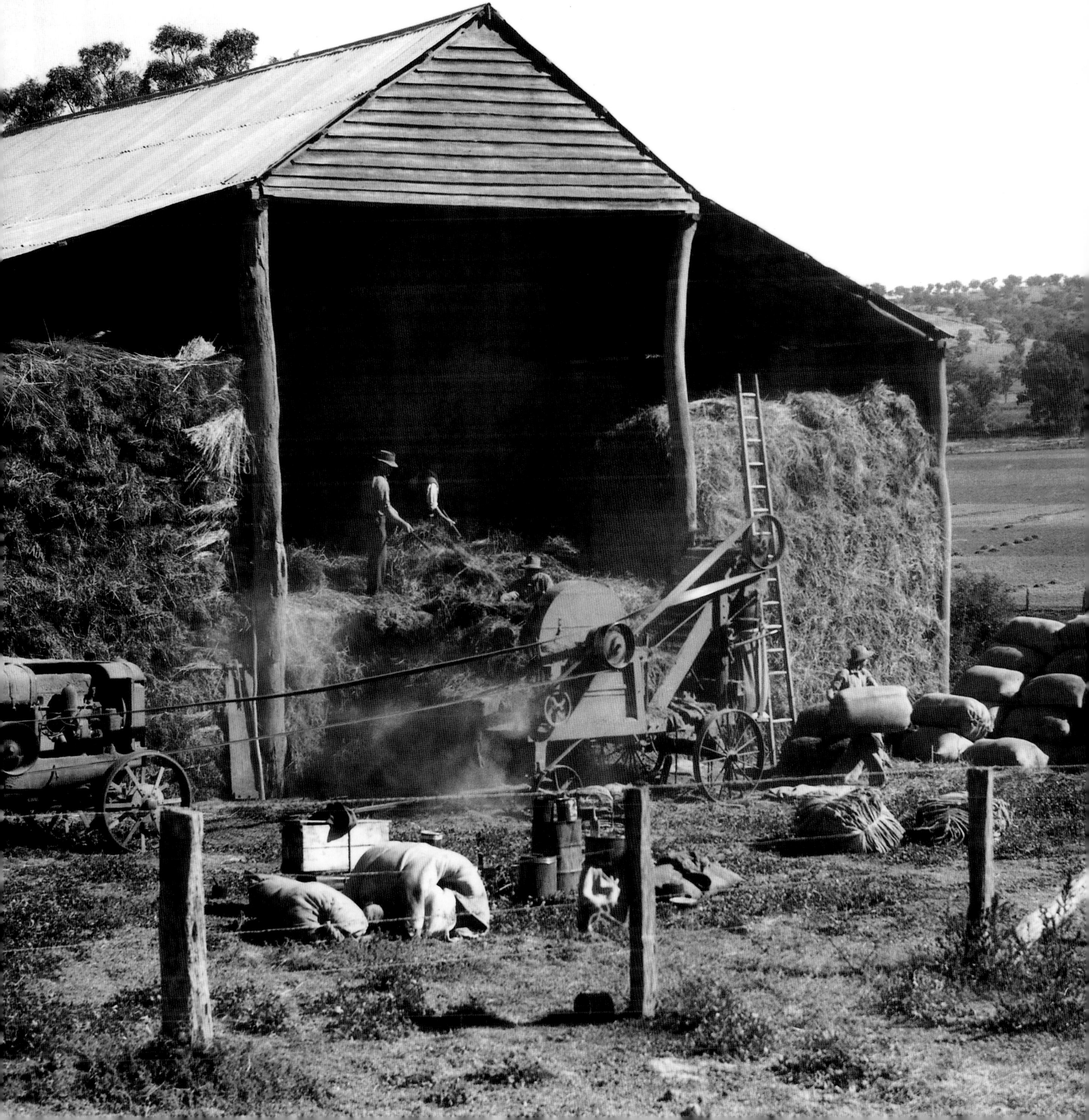

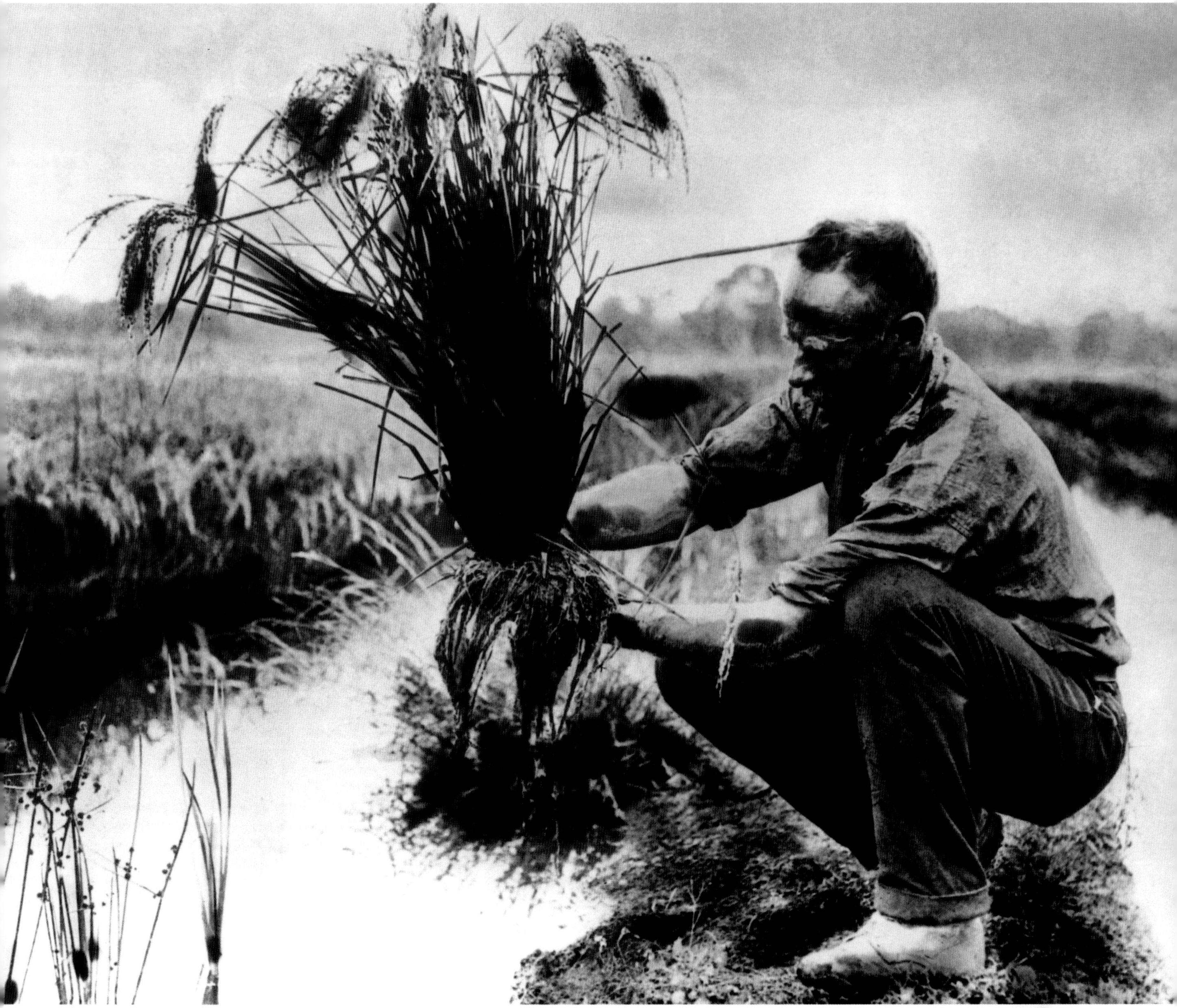

ABOVE: Many harvester owners operate as contractors and harvest crops around Australia, following the seasons.

LEFT: A rice paddy in the Riverina. Today the rice industry is a vertically integrated, market driven agribusiness producing around 1.4 million tonnes of high quality paddy rice. Rice growers in New South Wales produce the highest yielding rice crops in the world, averaging more than 9 tonnes a hectare.

SUSTAINING A NATION

ABOVE: Wheat storage at Temora, New South Wales, waiting for trains to carry the grain to ports or bulk handling centres.

RIGHT: A field of canola being grown in the south east of Australia. Oilseed crops are one of the success stories of Australian agriculture.

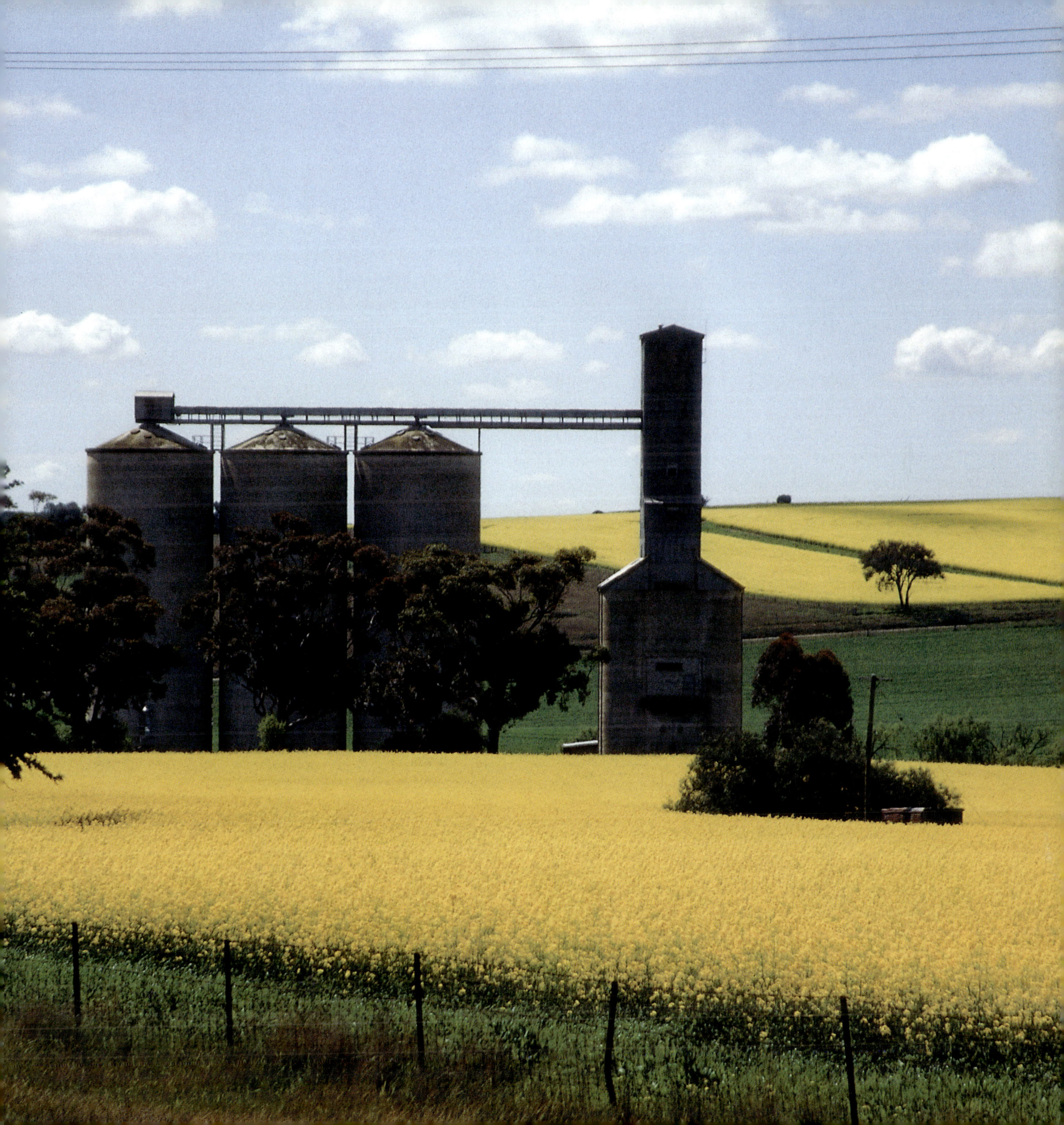

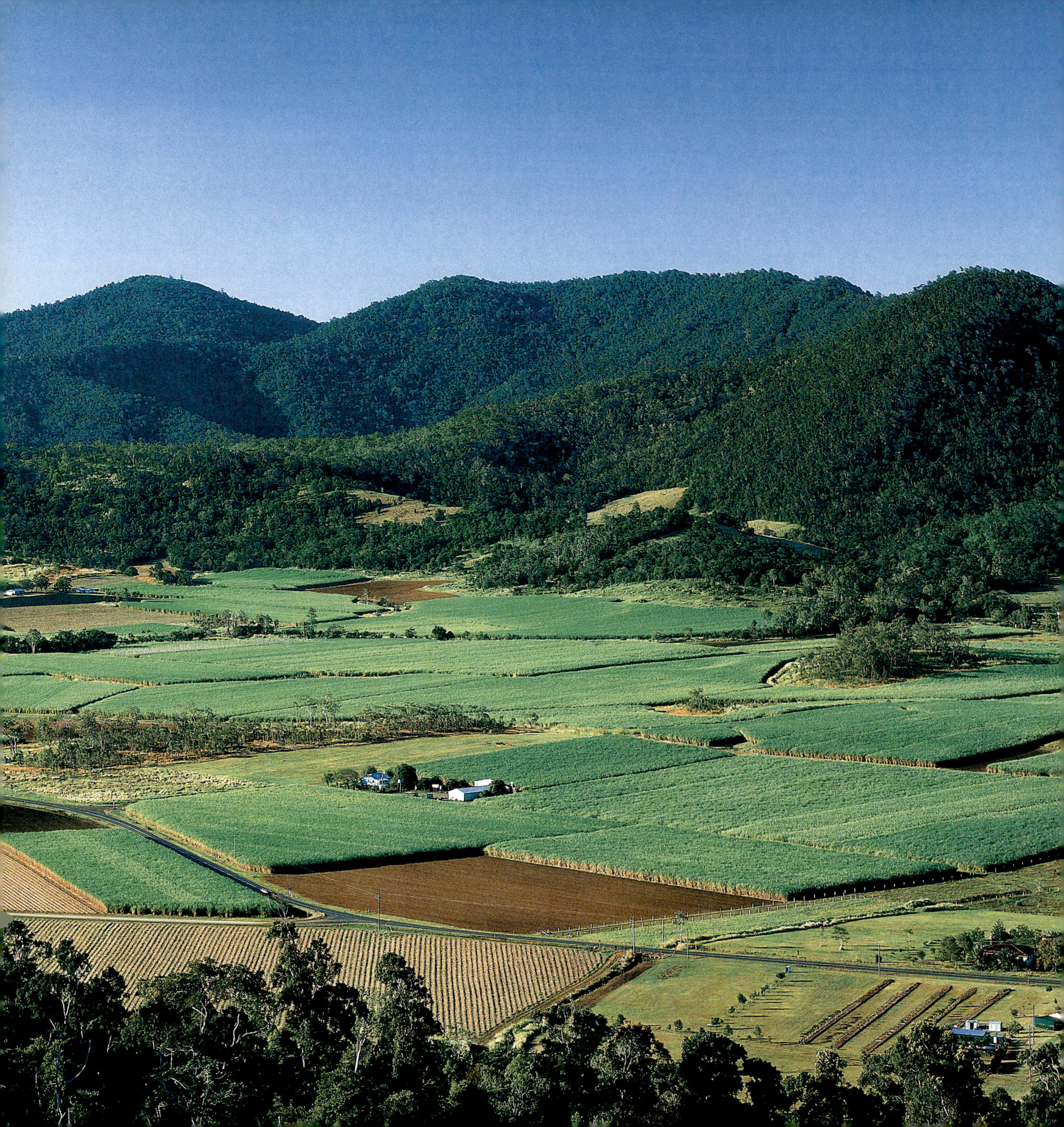

The Wealth of the Land

SUGAR

'The idea of carrying on the sugar industry by white labour only was in the nature of a huge experiment, and there were plenty of people shouting that it would fail, that no country in the world had ever attempted to grow cane by white labour and no country could.'

— Harry Easterby, *The Queensland Sugar Industry*, 1936

A COMMERCIAL SUGAR INDUSTRY organised on the plantation system and using indentured labor had existed in the coastal areas of Queensland and northern New South Wales since the 1870s. After Federation the industry was reorganised, as the new Commonwealth government prohibited the importation of Kanakas as part of its 'white Australia' policy in 1904. At the same time in a related development, the plantations were subdivided into small farm holdings (in the spirit of closer settlement) to produce 'white grown cane'. And to facilitate the establishment of a 'white grown' industry, the government paid growers a bounty from 1905 onwards.

The sugar industry was essential to the economy of tropical Queensland. Its very presence (indeed its expansion and one of the main reasons for its protection with a bounty) was that it was seen to be vital in terms of defending Australia's sparsely populated north from invasion. But the bounty was insufficient to compensate growers for the increased labor costs incurred when the Kanakas were replaced by a unionised workforce of cane cutters.

In 1915, the production and marketing of sugar was placed under the joint control of the Queensland and federal governments through a combination of state based legislation and agreements between the two governments that gave growers a degree of price stability. Under state legislation, farmers were allocated an acreage that was to come under cultivation, with the harvest assigned to a particular mill. The price payable by the mill to the growers was determined by an

OPPOSITE: Cane growing near Proserpine, Queensland.

This chapter is proudly brought to you by Sugar Research and Development Corporation *and* CANEGROWERS.

independent tribunal, the Central Cane Prices Board. Production quotas for the mills were also regulated. All raw sugar produced in Queensland was acquired by the Sugar Board, a statutory authority. The two refining companies, CSR Limited and the Millaquin Sugar Company, were contracted to refine and distribute the sugar, which was then sold to wholesalers at a fixed price. New South Wales participated in the scheme under a voluntary agreement. This degree of regulation remained in place, with few changes, until the 1980s.

However, there was no significant expansion of the sugar industry until the 1920s when the federal government increased the minimum price payable to growers. Then in 1923, the federal and Queensland governments negotiated the first of a series of sugar agreements. These provided financial security through a subsidy created by increasing the domestic price of sugar.

By now, greater efficiency in the industry had led to increased production, and Australian sugar was exported in 1924. But as exports were not subsidised, prices rarely exceeded those of domestic consumption, and the world sugar market was highly volatile. By the 1930s, the industry was in a desperate position. The 1937 International Sugar Agreement (which Australia signed) attempted to maintain prices and regulate competition using mechanisms such as export quotas. Ultimately the regulatory system, along with the sugar agreements, provided some basis for further expansion. These arrangements remained in place until the 1980s.

By 1938 more than half the industry's production was being exported to Britain. But it was not until 1951 that the industry enjoyed any real measure of stability, with the negotiation of the 21-year (British) Commonwealth Sugar Agreement, which allowed preferential access for sugar from Australia to British markets. Another important development was the United States Sugar Act which gave Australia a percentage of the US foreign sugar requirements. Faced with the expiry of these two longstanding agreements in the mid-1970s, the Australian industry sought (and found) new markets, principally in Asia.

While high labor costs were the bane of Australia's sugar industry, they also stimulated efforts, often by the growers themselves, to mechanise the backbreaking work of harvesting and transporting. Australia led the world with this technology and from the 1940s, harvesting machinery began to replace the itinerant cane cutters who had been such a prominent feature of our cane fields.

Today, Australia's efficient sugar industry leads the world in the adoption of technology and is a $1–2 billion contributor to the Australian economy with over 80 per cent of production exported to some fifteen countries.

ABOVE: No. 11 Loco with full cane bins heading for a mill in the Proserpine area.

OPPOSITE: Cane cutter in the 1930s. Though well paid, the arduous labor of cane cutting was only available during four or five months of each year. The demands of the job meant that the working life of the cane cutter was limited. Their itinerant and aggressively masculine lifestyle, was depicted in Ray Lawler's, *Summer of the Seventeenth Doll*.

SUSTAINING A NATION

RIGHT: Kanaka workers hoeing weeds in young cane in the early 20th century. The sugar industry in Queensland and northern New South Wales was established in the 1860s and 1870s largely using indentured labor from Vanuatu and the Solomon Islands.

LEFT: Italian cane cutters posing in the cane fields in 1930. In the 1920s, there was an influx of southern Italians to Australia, many finding their way to the expanding cane fields of north Queensland.

LEFT: Harvesting using a portable rail line. The laborers on the left of this photograph are Kanakas; the small group on the right harvesting the sugar cane are Chinese or 'coolie' laborers. During the first decade of the 20th century, legislation was enacted to remove colored labor from the sugar industry via an excise Act and a bounty on 'white grown sugar'.

RIGHT: Cane train crossing Mulgrave River. Replacing the horse, mill owned or government owned trains meant that cane, which deteriorates rapidly after cutting, could be transported to the mills more quickly.

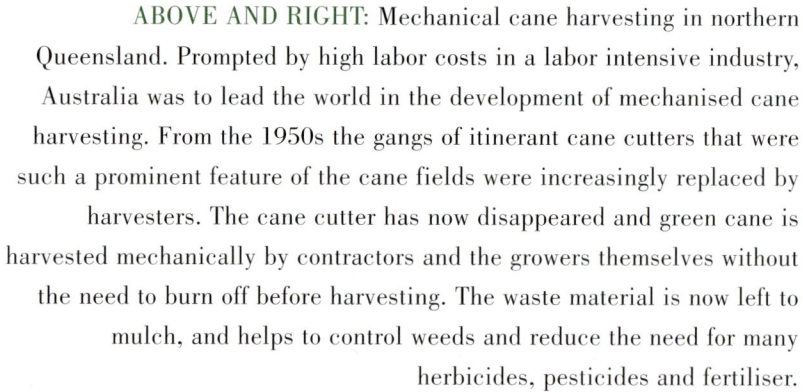

ABOVE AND RIGHT: Mechanical cane harvesting in northern Queensland. Prompted by high labor costs in a labor intensive industry, Australia was to lead the world in the development of mechanised cane harvesting. From the 1950s the gangs of itinerant cane cutters that were such a prominent feature of the cane fields were increasingly replaced by harvesters. The cane cutter has now disappeared and green cane is harvested mechanically by contractors and the growers themselves without the need to burn off before harvesting. The waste material is now left to mulch, and helps to control weeds and reduce the need for many herbicides, pesticides and fertiliser.

LEFT: Cane receival at Maryborough Mill in the late 19th century. Sugar trains would eventually replace the horse in carting cane to the mills.

BELOW LEFT: Roller milling plant. Steam engines were used for driving these horizontal mills. Canes were fed from a table, and after being squeezed between the first and top under roller were guided between the top and second under roller by a plate called the 'dumb turner'.

BELOW: Modern milling plant receiving harvested cane.

SUSTAINING A NATION

49

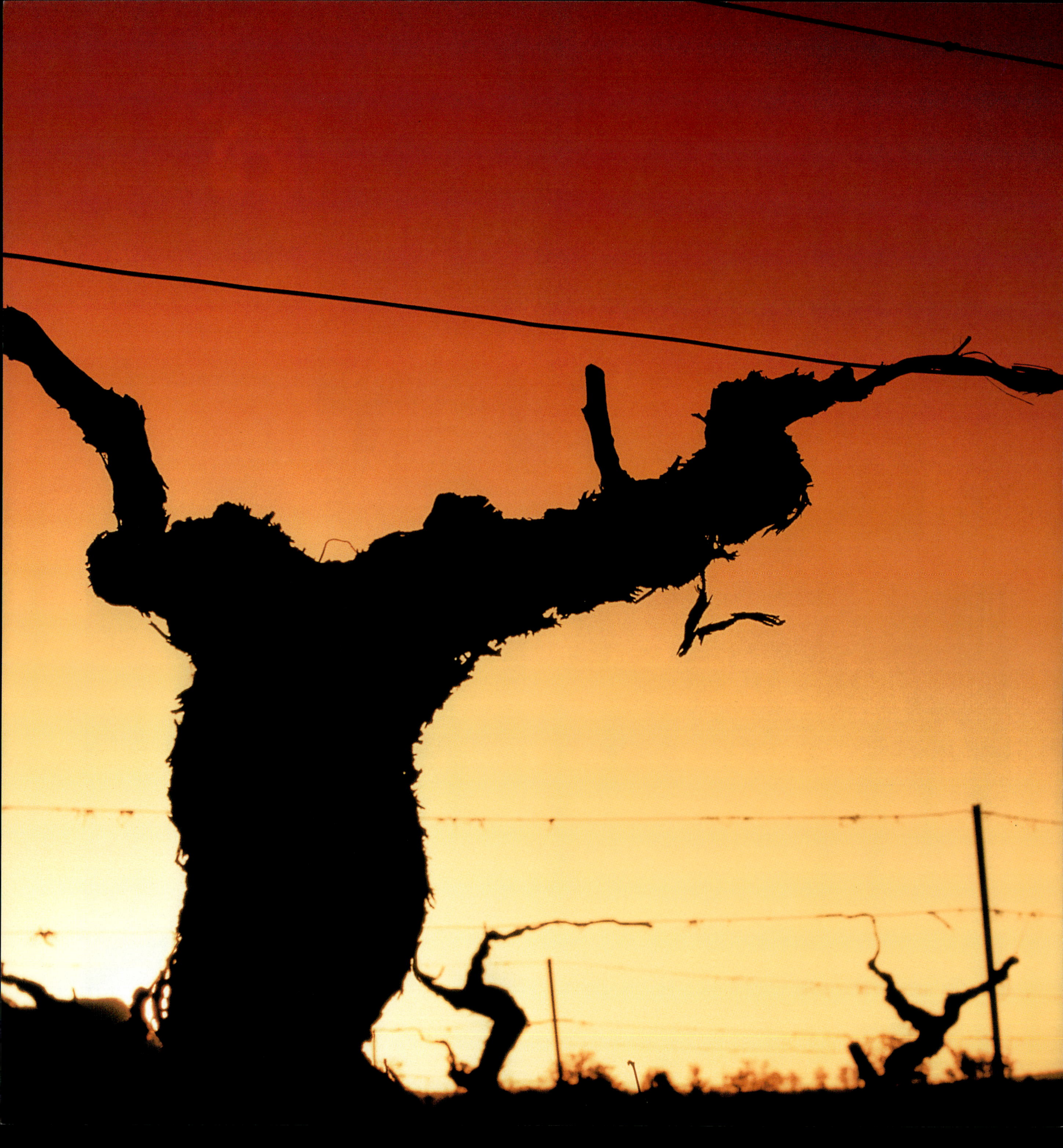

The Wealth of the Land

WINE

'Were I asked to name the one industry on which the prosperity of Australia must sooner or later rest, I should unhesitatingly answer, "On the cultivation of the vine". And this must be so; for while there is every reason to know that it will be called for from abroad, it is absolutely certain that it will be required in our own territories.'

— Phillip E Muskett, *The Art of Living in Australia*, 1893

THE MODERATE DRINKING of 'light table wines' was recognised for its medicinal benefits for anaemic or sickly patients in the Australian colonies in the 19th century. Wine was also lauded as a temperance beverage — a healthy alternative to rum.

At the beginning of the 20th century, although vines had been planted right across the southern half of Australia and there was a thriving export industry, wine continued to be 'almost a curiosity' on the tables of Australian households.

With the opening of the irrigation areas along the Murrumbidgee and Murray rivers and the arrival of the first Italian migrants after the First World War, production of cheap fortified wines became the industry's driving force. In the 1920s, exports of fortified wine to Europe reached levels not seen again until the 1980s. Fortified wines remained the focus of the industry until the mid-1950s; there was no sizable domestic market for table wine until the 1960s.

After Federation, the commercially astute South Australian industry quickly became the country's leading wine producer. Removing the interstate trade barriers had allowed wines from the Barossa, Eden and Clare valleys and McLaren Vale to establish a strong foothold in Victoria, New South Wales and Western Australia.

South Australia's success led to a substantial contraction of New South Wales wine production. The Hunter Valley — a commercial wine making region since the 1850s — maintained its place as a premium wine producing region through noted

OPPOSITE: The First Fleet brought seed of the claret grape and several rooted vines that were planted in the Governor's garden soon after arrival.

ABOVE: Pruning vines at Dookie Agricultural College, Victoria in the 1920s. While Victoria was the leading producer of wine in the 1890s, the devastation of its vineyards by *phylloxera* saw the industry fall into decline by the turn of the century, only to re-emerge by the 1960s.

wine makers such as Dan Tyrrell, Hector Tulloch and Maurice O'Shea. But by 1956 the vineyard area had diminished to a mere 466 hectares. Other wine areas such as Mudgee — where only Craigmoor (formerly Rothview) remained in production — also fell into decline until the 1970s.

Victoria had been the leading wine producer in the 1890s. But its vineyards had been devastated by *phylloxera*, which first appeared in the 1870s, and was still in evidence until 1915. By the 1950s only a few vineyards, including Chateau Tahbilk, Great Western and Rutherglen, remained.

The Swan Valley produced most of Western Australia's wine and its output was boosted in 1918 by the arrival of Yugoslav migrants. Though Jack Mann's white burgundy for Houghton in the 1940s was the first of its style in Australia, the industry here also stagnated.

The wine industry's turnaround came with the prosperous 1960s. Australian's leisure patterns were changing; dining out, once the preserve of the wealthy few, became more common and people began to enjoy a glass of wine with their meal. At the same time European migrants were having a profound influence on the restaurant menu.

The regeneration of the Victorian wine industry was by now under way in the Yarra Valley. Vineyards were also being re-established in New South Wales' Hunter region. By the 1970s Australians had taken to wine with such enthusiasm that it seemed to be a feature of almost every social gathering. From the mid-1980s Australians finally moved on from the carafe and the cask to drinking premium bottled wines — 'drinking less but drinking better'.

With the regeneration of the wine industry came an expansion in exports as Australian wines made their mark internationally. In 1985 exports amounted to around 8 million litres (worth $21 million); by 1999 this was over 230 million litres (worth around $1.15 billion). It is estimated that this will jump to over 500 million litres by 2002–03.

The vast size of Australia, the diversity of its climate and soils, the absence of pollution and the generally low urban pressure all contribute to Australia's unique array of wine styles. And underlying this are wine making skills honed to a fine edge by a combination of world class research and education facilities and the typically pragmatic and committed nature of Australian wine makers.

There have been important spinoffs from the recent development of the Australian wine industry. Australia has led the world in developing mechanical pruning. Viticultural research has resulted in profound changes in the way grapes are grown, in the varietal composition of Australia's vineyards and in the areas in which grapes are grown. The Australian Wine Research Institute, established in 1954, is world renowned, and our technological innovations and research are now an export commodity in their own right.

ABOVE: Crushing grapes in the Horsley Winery in New South Wales in the late 1890s.

SUSTAINING A NATION

53

ABOVE: A Penfold's Wines display in the windows of Fred Dineen's Kings Cross 'wine café' in the late 1930s.

RIGHT: A city wine bar of the same period promoting Penfold's Wines. With the abolition of trade barriers after Federation, South Australia established itself as the country's leading wine producer. Penfolds opened an office in Sydney in 1904. Along with hotels, wine bars were a target of the temperance movement and subject to restrictive liquor legislation in most states from around 1910. Until the 1960s, their numbers and trading hours were strictly regulated and policed.

SUSTAINING A NATION

ABOVE: Pickers depositing grapes on the back of a dray in the late 1930s.

LEFT: A load of mechanically harvested grapes. There has been a radical shift from manual labor to mechanisation in virtually every aspect of grape growing and wine production.

SUSTAINING A NATION

ABOVE: Mechanical harvesting in the Hunter Valley, New South Wales. Australia now machine harvests over 90 per cent of its annual crush.

LEFT: An Italian family harvesting grapes in the late 1920s. At this time there was an influx of European migrants to Australia because of immigration restrictions into the United States. Many Italians established vineyards in the Murray and Murrumbidgee irrigation areas. Along with German settlers who opened up the Barossa Valley in the 1840s, migrants have played a significant role in the development of the wine industry.

ABOVE: Wine casks being seasoned in front of Chateau Tanunda's distillery in South Australia in about 1910.

RIGHT: Barrel maintenance after 'racking' of red wine at Tyrell's Winery in the Hunter Valley, New South Wales.

SUSTAINING A NATION

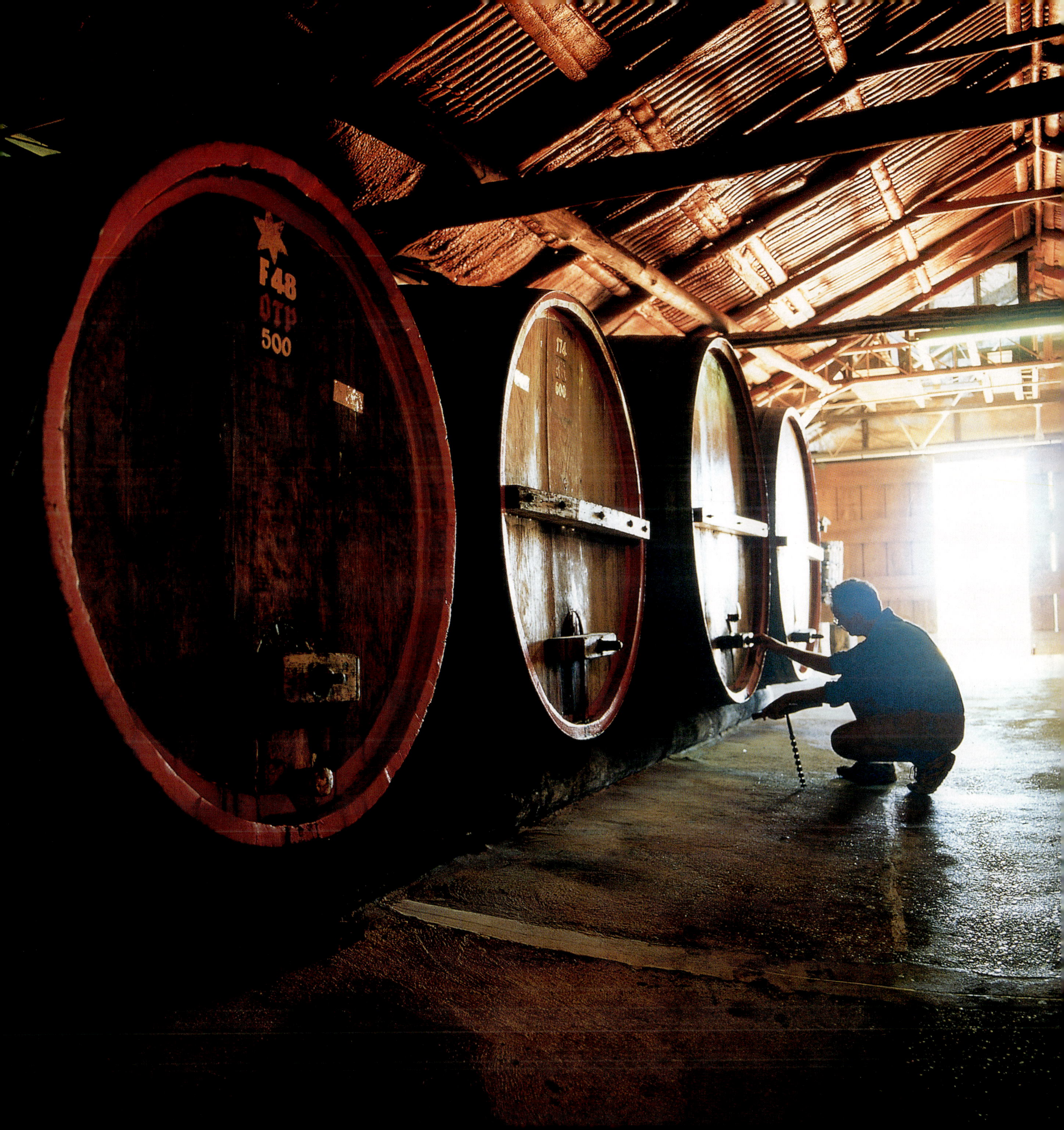
F48
500

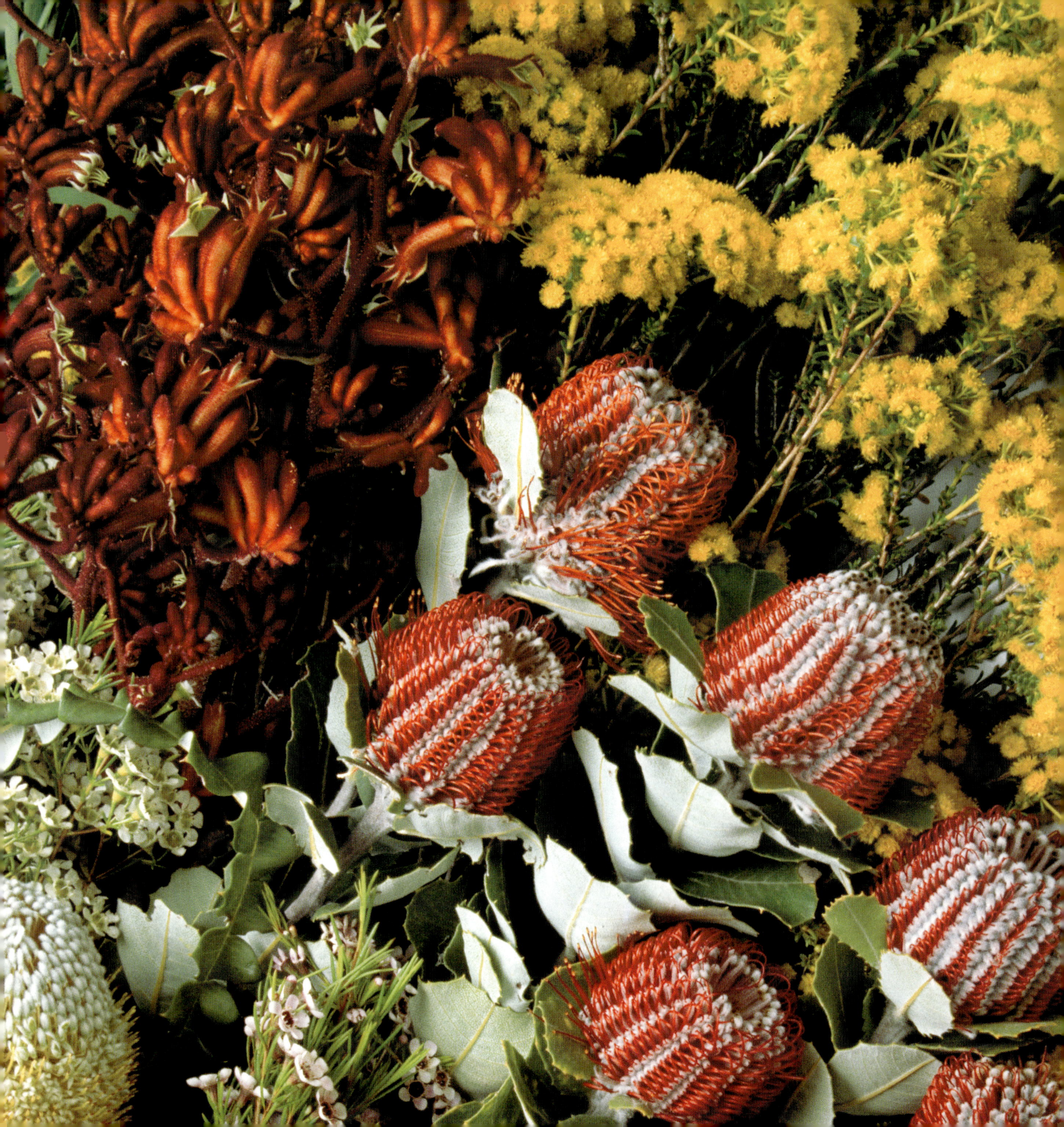

The Wealth of the Land

HORTICULTURE

'Most visitors to Australia exclaim at the abundance and variety of fruit which is offered for sale in the shops. Large juicy peaches can be obtained, during the summer, at a penny each or even less, while the best grapes rarely cost more than sixpence a pound.'

— GS Browne, *Australia: A General Account*, 1929

AUSTRALIA has one of the most diverse horticultural industries in the world, much of which it owes to our multicultural mix and our wide range of climatic zones. British settlers brought staples such as apples, pears, potatoes, pumpkin, carrots and the brassicas. The southern Europeans planted tomatoes, eggplant, fennel, zucchini, artichoke and a range of other herbs and vegetables, and they brought olive oil. More recently, migrants from Asia have introduced bok choy and a wide variety of tropical fruits.

By the late 19th century the types of horticultural production best suited to the individual colonies was established. As GS Browne observed in 1929: 'The wide range of fruit available is due to the fact that the supply comes from far-scattered States and districts differing very much in climate and soil. Take for example, the display in the windows of a Sydney fruit store in January. There will probably be bananas, pineapples, and mangoes from Queensland; apricots and strawberries from Tasmania; and plums, nectarines, peaches and late cherries from the middle States. Later in the year will come apples and pears, grapes, and finally oranges in large quantities.'

From the 1880s horticulture expanded considerably, largely thanks to the Chinese market gardener who showed what a difference irrigation makes on a dry continent with variable rainfall. The first irrigation settlements were on the Murray River, at Renmark in South Australia and Mildura

OPPOSITE: Most Australian governments — state and federal — had enacted legislation for the protection of wildflowers by the 1930s. Today, Western Australia is the nation's leading exporter of cutflowers and foliages. While traditional flowers dominate the domestic cutflower market, the export market is dominated by native flora and foliages.

This chapter is proudly brought to you by Sydney Markets Limited.

RIGHT: Drying sultanas on hessian bags in the Riverland, South Australia. Once dried, the sultanas were then boxed in large shallow trays (called 'sweats') for dispatch to the packing houses. These vine fruits were largely responsible for the growth in dried fruits. But, plagued by the difficulty of competing in world markets, the dried fruit industry remained reliant on government subsidies and import restrictions for its viability.

OPPOSITE: A soldier on leave during the Second World War harvesting carrots. As with commodities and items necessary to rural production, labor shortages were so acute during the war that from 1943 experienced farm workers were granted leave to go home for the harvest.

in Victoria — known as the Riverland and Sunraysia regions — where the citrus industry (among others) was established. The construction of the Burrinjuck Dam (1912–15) on the Murrumbidgee led to irrigation settlements for fruit growing at Leeton and Griffith.

In the 1920s and 1930s there was a proliferation of major irrigation schemes along the Murray, Goulburn, Macalister, Lachlan, Edward and Collie rivers.

There was great optimism about the possible markets for the fruits of such intensive cultivation, and during hard times, it was believed, the versatility of the irrigation farmer would prevail. But the Depression proved otherwise — irrigated crops suffered the same market drop as other agricultural produce.

The area under irrigation increased rapidly after the First and Second World Wars under the soldier settlement schemes. These 'fruit salad' blocks grew 'a bit of everything', including grapes, stone fruits and citrus. But the industry was not without its problems. Small holdings and the labor intensive nature of horticulture limited both the size of production and profitability.

The advent of canning and drying techniques, and of cool storage and refrigeration, provided a great boost to fruit growing. Fruit canning and jam making began in Tasmania. Commercial jam making had been started in Hobart in 1861 by

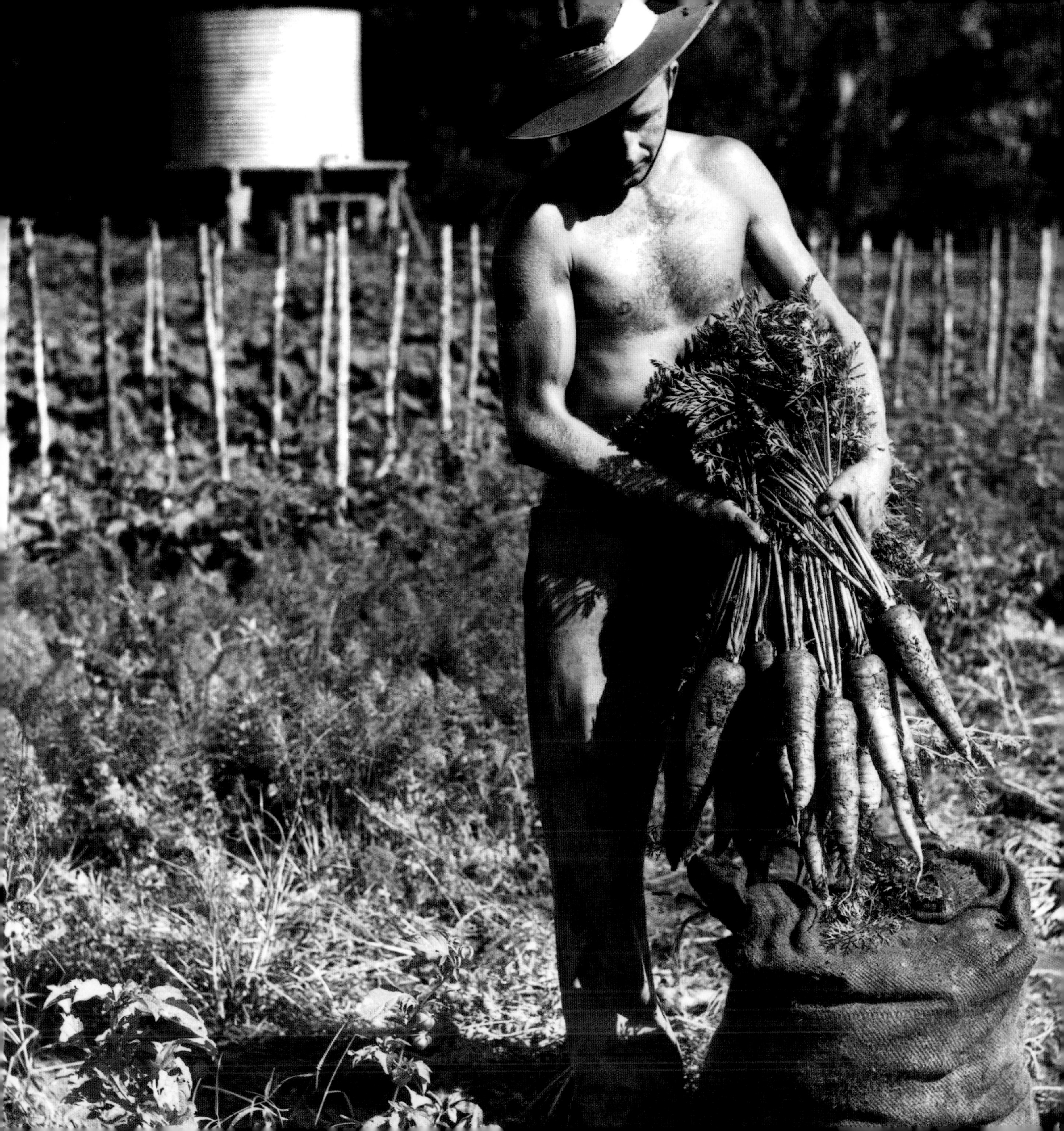

ABOVE: Irrigation channel in 1908 at Rochester, Victoria, in the Murray Valley in the infant days of water conservation. Irrigation, the new weapon against drought, was central to the development of horticultural production. But lack of experience and expert advice in the early days often resulted in serious problems, with waterlogging and salinity (not to mention lack of water) causing destruction of plants.

George Peacock, and Sir Henry Jones began his canning operation, IXL, in 1891. In 1910 the first automated assembly line for filling, cooking and closing the cans came into operation. Canneries were set up in all major growing areas and became an important source of employment for rural communities.

Chinese market gardeners dominated commercial vegetable growing from the gold rush days until 1901 when immigration restrictions led to a decline in their numbers. Market gardening continued to be the principal form of vegetable production until improved transport, together with encroaching urban development, caused a shift to farming areas.

Market gardening in outer metropolitan areas, however, continues to be an important source of horticultural production. Unlike the irrigation areas, where large scale, specialised growing is now the rule, the market garden is still a small scale enterprise. And as the European migrants took over from the Chinese — and South East Asian migrants from them — it has come to play a significant role in introducing Australians to new vegetables.

The first vegetables commercially processed (in very small quantities) were tomatoes and asparagus. In 1912, a handful of vegetable canning plants was established for processing asparagus, peas and a few root vegetables. After the First World War, sweet corn, beans and mushrooms were added to the list. Unlike fruit processing, vegetable processing was limited until the Second World War when production was increased in all states to supply the forces. Today, vegetable processing is a significant commercial activity.

The growth in consumption of fruit and vegetables in the last hundred years and the changes in the types we eat reflect an increased awareness of their importance in our diet and the changing cultural influences within our society.

CUTFLOWERS are a growing export industry for Australia. The market, dominated by wildflowers and foliages native to Western Australia, also includes South African proteas, and dried and preserved flowers.

RIGHT: Harvesting potatoes in the 1930s. Potatoes have been grown in Australia since 1788. Marketing methods have evolved from simple arrangements through merchants from the latter part of the 19th century, to the Australian Committee (which controlled production and distribution during the Second World War), to the producer controlled marketing boards which were set up immediately after the war.

LEFT: Grading and packing apples at Karragullen, Western Australia, in 1933. The export of apples began with the first trial shipment to England from Tasmania in 1884. The successful delivery soon led to a rapid increase in their production for British markets in Tasmania, followed by Victoria and later in other states.

SUSTAINING A NATION

LEFT: Tobacco leaf strung on poles in preparation for drying in the sun in 1904. This method was introduced by Chinese farmers in the 19th century. Queensland became the major tobacco producing state in the first two decades of the 20th century. The Chinese workers pictured here were replaced by British migrants with the introduction of the 'white Australia' policy. Italians and a much smaller number of Chinese workers were to provide the labor for tobacco growing from around 1930.

RIGHT: A harvest of Queensland's famous 'blue' pumpkin. Once despised as cattle fodder in some European countries, pumpkin has become both a popular food and a major contributor to vegetable production during the 20th century. In fact, it enjoyed a brief period of fame in the early 1980s when Flo Bjelke Petersen, the wife of the then Queensland premier, published her famous pumpkin scone recipe. She later became a senator.

SUSTAINING A NATION

ABOVE: Land Army girls, Marge Craig and and Jen Stevenson, in the poppy fields at the Dickson Experimental Station in the Australian Capital Territory around 1943–44. As part of the war effort, official photographers visited both large stations and small farms to document the activities of those who were 'helping to keep things going' — in this case growing opium poppies for morphine production during the war. Poppies are grown in Tasmania today under strict supervision.

SUSTAINING A NATION

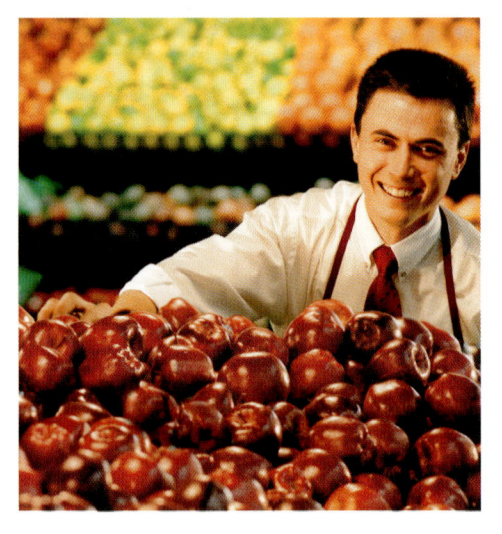

ABOVE AND LEFT: Display has long been important as these pictures of a 1930s' fruit stall (above) and a Woolworths Supermarket 'Fresh Food People' advertisement (left) show.

RIGHT: Australia's first commercial vegetable growers were the Chinese who planted gardens during the gold rush to sell to the miners. Later Bendigo's Spanish community founded the tomato industry in Victoria using their traditional gravity irrigation methods. During the 20th century, improvements in transport and the emergence of supermarkets saw a shift from mixed market gardening to growing specialised crops. Today some suppliers, such as these celery growers, produce fruit and vegetables for a specific company — in this case Woolworths.

SUSTAINING A NATION

SUSTAINING A NATION 69

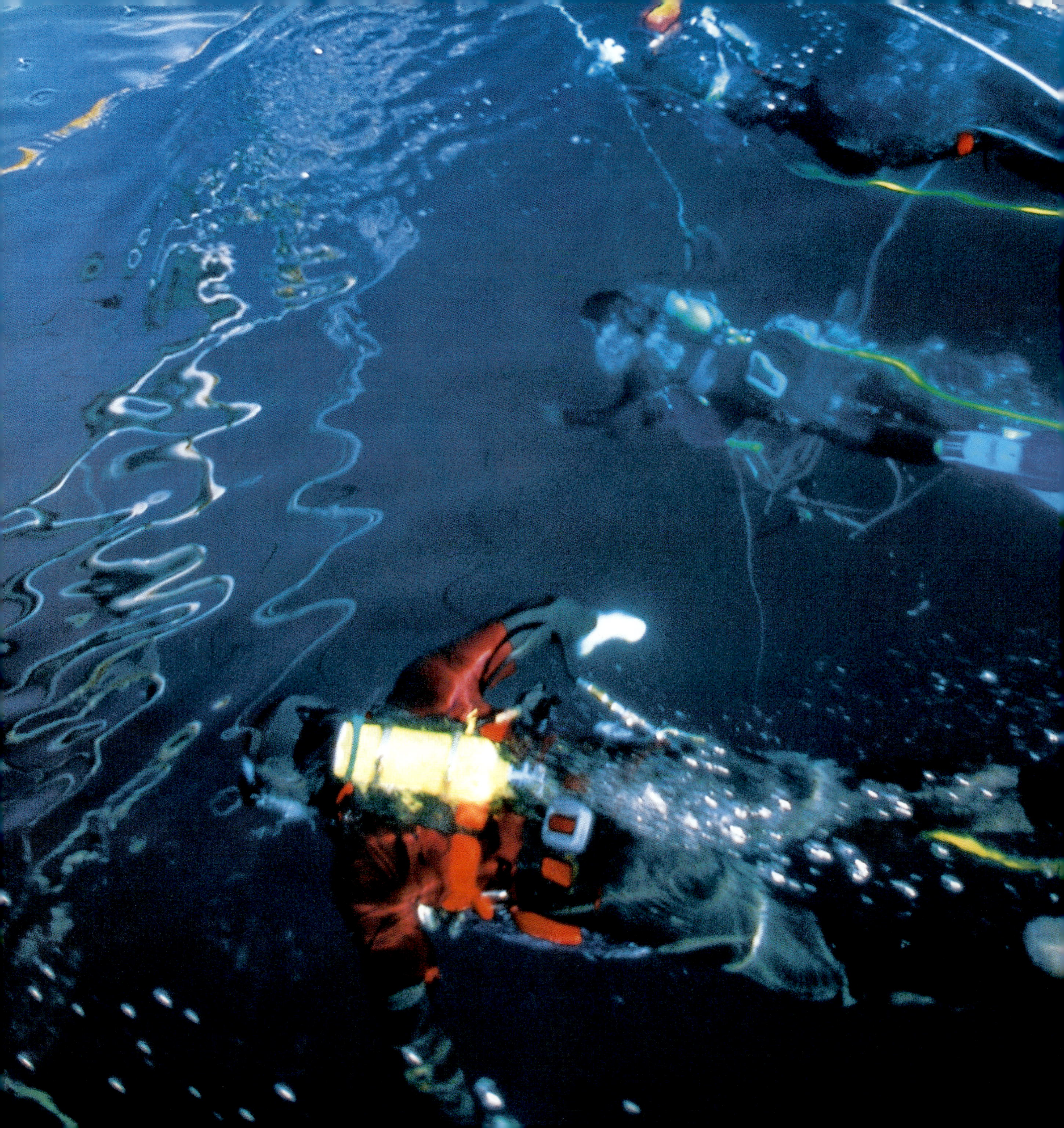

The Wealth of the Land

FISHERIES

'Up until 40 odd years ago Catholics did not eat meat on Friday because that was the day Christ had died. Because it was a compulsory meal on Fridays, few of us liked it then. We had to eat it, but we didn't have to like it. I noticed that people who gave the church away avoided eating fish for many years afterwards. Only gradually did they discover that there was a wide and tasty variety of fish in Australia, once you had the choice.'

— Edmund Campion, 2000

THE MOST FAMOUS of all Australian seafoods is probably the prawn. It entered the vernacular in the 1940s with 'don't come the raw prawn with me', by the 1970s it was a sign of dining sophistication as a prawn cocktail, and in the 1980s it gained an international reputation when Paul Hogan convinced Americans that 'throwing a shrimp on the barbie' was our national pastime.

Crustaceans (prawns, lobster, bugs, scampi and crab) and molluscs (abalone, oysters and squid) are our prized jewels of the sea, and they help account for the doubling of seafood consumption in Australia since the 1950s. The export of prawns to the United States, which began in 1955, has led to prawn fisheries being established in Western Australia, South Australia and the Northern Territory.

PEARLING One of Australia's earliest commercial fishing industries was pearling. It began in northern Australia in 1861, and quickly became the basis of the economy of many coastal towns. Pearl shells were also sought for their mother-of-pearl. At one time, Australia produced around 80 per cent of the world's mother-of-pearl.

Broome was already the most important pearling area by the 1890s, but by the late 1930s the seabeds were exhausted. Cultured pearl farms were established in the 1950s when the Japanese, who had played an important part in the pearling industry from the late 19th century until the Second World War, once again became involved.

FISHING In 1915 the New South Wales government established a small fleet of steam trawlers to fish

OPPOSITE: The first pearl divers were Aborigines and Torres Strait Islanders. Many women divers were forced into prostitution leading the Western Australian government to prohibit indigenous women from the proximity of pearling boats. Timorese and Javanese divers were recruited after that.
(Photograph © Richard Smyth)

This chapter is proudly brought to you by AFFA, AFMA, ASIC *and* Sydney Fish Market Pty Limited.

SUSTAINING A NATION

ABOVE: Having largely disappeared by the late 19th century, whaling re-emerged as an important industry in the 1950s with stations established at Albany (1952) and Byron Bay (1954). Mounting opposition prompted an independent inquiry in 1978 and the passing of the Whale Protection Act in 1980.

the east coast grounds (from Port Stephens south to Bass Strait). Big catches were made, and although the hauls remained good, the fleet fell into financial difficulties and was sold to private interests in 1923. The price of profitability was depletion of stocks and falling catches and in the late 1940s the steam trawlers withdrew from the east coast.

By the 1930s Danish seine fishing was being practised in New South Wales. The catches were exceptionally good for a few years, but again the grounds were overfished, and by 1941 many of the boats were finding the industry unprofitable.

Today, deepwater trawling by Australian fleets is generally confined to waters off New South Wales, Victoria and Tasmania, and the principal catch includes morwong, redfish, orange roughy, school whiting, flathead and snapper.

The pelagic resources — fish that swim in or near the surface of the water, such as pilchards, mackerel and tuna — were largely untapped until the 1950s, and are now fished from Albany and Esperance in Western Australia and Port Lincoln in South Australia.

While Italian and Yugoslav migrants figured prominently in the development of Australia's commercial fisheries from the 1920s, Greek migrants became involved in distribution, particularly in establishing fish and chip shops in suburban and rural areas.

Traditionally, Australians of British and Irish stock were not big fish eaters — except on Fridays (Catholics were compelled to eat fish on Fridays until the mid-1960s). Those who lived inland, where the freshness of fish was suspect, ate canned fish transformed into tuna mornay, salmon 'curry' or salmon rissoles instead.

However, as fresh fish consumption increases, so too do concerns about depletion of stocks. Australia's commercial fisheries are mainly confined to a relatively narrow continental shelf, and it seems there will always be a shortage of fresh fish from Australian waters; marine resources are not inexhaustible. Federal and state governments have introduced measures aimed at conserving these resources and sustaining the industry, including catch quotas, restrictions on the number of boats, closed seasons, closed areas and minimum size. But, illegal foreign fishing is a major issue. The exclusive fishing zone of 200 nautical miles has imposed certain rights and responsibilities on Australia — the right of first access to fisheries and other resources in that area, plus the responsibility for determining the maximum sustainable yield. The 1993 Convention for the Conservation of the Southern Bluefin Tuna formalised an agreement between Australia, Japan and New Zealand for global management of this species, which has suffered serious depletion.

Fishing remains an important commercial enterprise providing a livelihood for thousands of Australians. It is also a recreational hobby for some four million Australians who simply enjoy 'dangling a line'. And today eating fish regularly is widely acknowledged to be essential for our health and well being.

AQUACULTURE Aquaculture has developed significantly since the 1980s and now includes marine and freshwater finfish, crustaceans and shellfish. The value of acquaculture production is now around $600 million, about 30 per cent of the total value of Australia's fishing industry.

ABOVE: Tuna farming in South Australian waters.

SUSTAINING A NATION 73

LEFT: Displaying the local catch of Murray Cod. While freshwater fish make up a small proportion of the total Australian catch, rivers provided an important source of fresh fish for inland populations. The fish population in the Murray and its tributaries has been declining since the 1950s for a number of reasons including overfishing, silting of the river and muddying of streams because of land erosion, damming, and competition by introduced species such as carp.

SUSTAINING A NATION

ABOVE: Eden, the main port on the New South Wales south coast, was originally a base for the whaling industry. From 1903, Eden was an important port for timber ships servicing the nearby sawmills. Fishing is the major industry that has survived from earlier times and today's catch, destined for the fish markets in Melbourne and Sydney, includes tuna, snapper, morwong, flathead, redfish, gemfish, salmon and abalone.

RIGHT: A good haul of red spot whiting and red mullet.

SUSTAINING A NATION 75

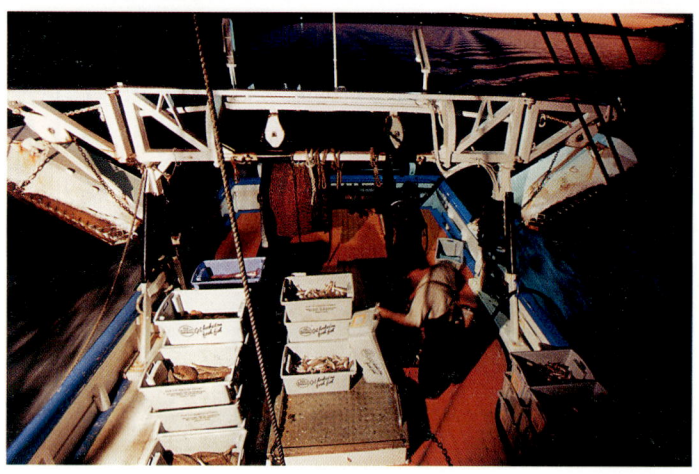

ABOVE AND LEFT: Australian fisheries operate in one of the world's cleanest environments to supply a wide range of seafood products to domestic and export markets. The gross value of seafood production has broken the $2 billion barrier. Today, there's a cooperative partnership between fisheries managers, commercial operators, scientists, environmentalists and recreational fishers to sustainably manage fisheries, aquaculture and to protect the marine environment.

RIGHT: Today the pearling industry is dominated by the Paspaley and Kailis families, who produce pearls of outstanding size and lustre in the pollution free waters of Northern Australia. (Photograph © Richard Smyth)

SUSTAINING A NATION

SUSTAINING A NATION

77

LEFT: Fishing provides a livelihood for thousands of Australians. It is estimated today that some 30,000 people are directly employed in the fishing industry, with many more involved in processing, retail and restaurants.

RIGHT: The Sydney Fish Market's auction floor where more than 70 tonnes of product changes hands on a typical trading day. The market uses an adapted version of the Dutch auction system (originally devised to sell flowers efficiently) to sell about 90 per cent of product. The sale process is fully automated with some 170 buyers bidding via computer terminals. For live product such as mud crabs, the traditional auction system has been retained, and for high-grade sashimi there is a special buying pavilion.

SUSTAINING A NATION

PRICE
UNITS
BUYER
CRATES
REMAINING
MAXIMUM
F.A.S.T Auction System
KEL AEROSPACE
Lot
Posn
Crates
Kgs
Supplier
Species Description
NEXT LOTS FOR SALE
No Eating No Smoking No Drinking ON AUCTION FLOOR

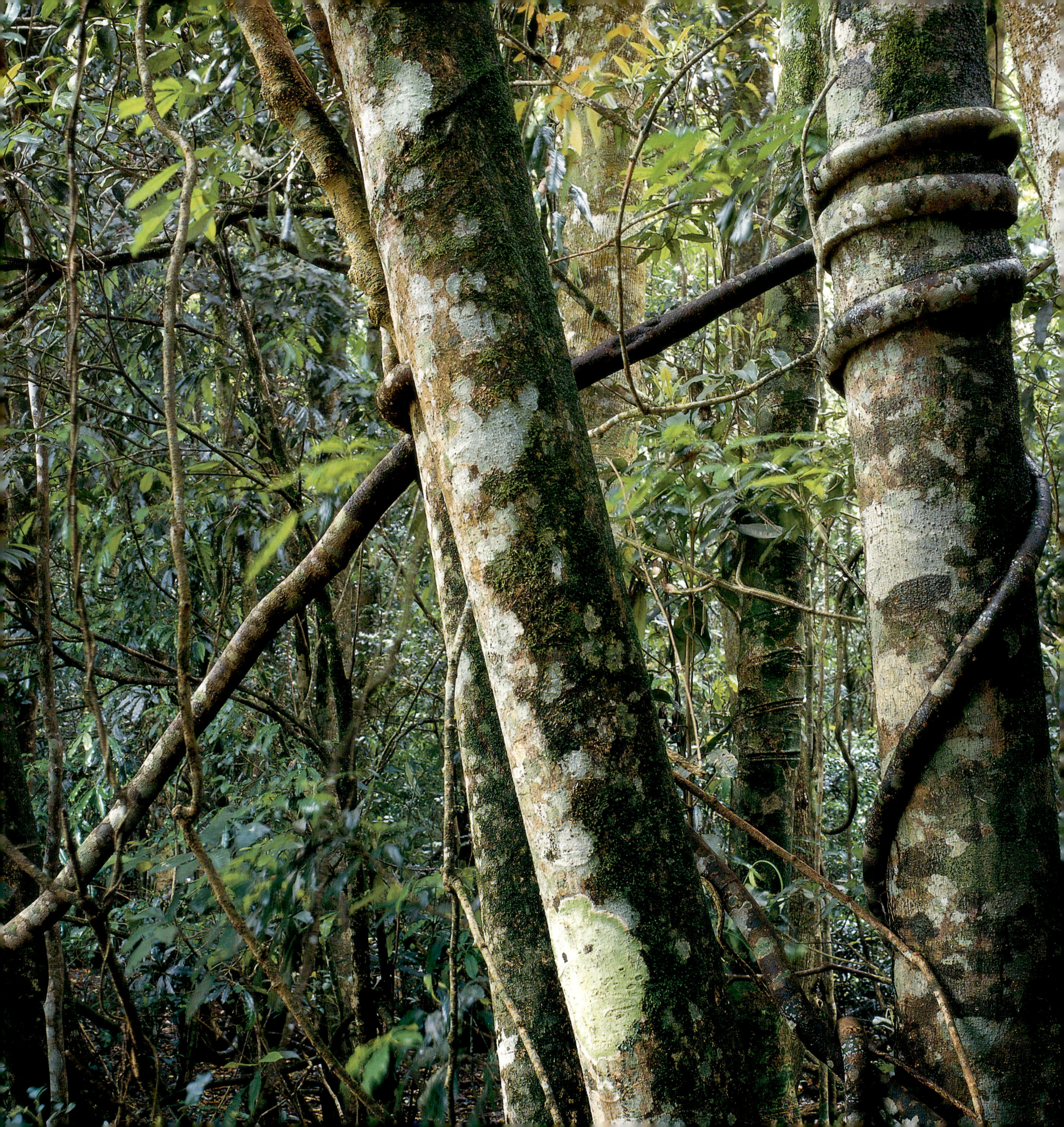

The Wealth of the Land

FORESTRY

'Australia is, with very trivial exception, one huge, unreclaimed forest … The choked-up valleys, dense with scrub and rank grass and weeds, and the equally rank vegetation of swamps, cannot lend to health. All these evils, the axe and the plough, and the fire of the settlers, will gradually and eventually remove; and when it is done here, I do not believe that there will be a more healthy country on the globe.'

— W Howitt, *Land and Labour and Gold*, 1855

IN THE YEARS FOLLOWING FEDERATION, the green and gold of the eucalpyt and wattle became Australia's national colors. Their elevation to national symbols also marked an important shift in our attitude toward the natural environment. From the late 19th century, the 'Australian bush' came to be regarded with growing affection.

For the early settlers in the Australian colonies, trees were a nuisance, to be cleared to make way for the higher purpose of agriculture. However, the commercial potential of the native timbers such as red cedar and maple was quickly recognised, and a thriving export industry grew up, particularly in the more exotic timbers. Mining and railway construction also increased the demand for timber, as did the traditional domestic purposes — houses, furniture and fuel. And with such vast quantities of forest available, 'few were yet aware of the desirability of conserving Australia's timber as a renewable economic resource; fewer still were convinced of the ecological or aesthetic arguments for restraint with the axe', noted historian Geoffrey Bolton.

By the end of the 19th century there were calls for better forestry management. South Australia was the first to acknowledge the need for a coherent forest policy — it created a woods and forest department in 1883 to manage the state's forest resources and establish new plantations.

Commissions of inquiry in Tasmania and Victoria in 1898 and in Western Australia in 1903 concluded that forests needed to be managed in a

OPPOSITE: Dorrigo National Park. Rain forests were harvested for high quality cabinet timbers such as northern silky oak, red cedar, Queensland maple and rosewood. Indigenous softwoods such as the hoop pine of northern New South Wales and Queensland were equally prized.

This chapter is proudly brought to you by NSW Forest Products Association Ltd.

closely by Victoria's Ferntree Gully National Park, near Melbourne, in 1882.

But it was not until the 1890s, when large sawmills were built in Western Australia and later in Tasmania, that significant areas of forest started to be reserved. In 1920 an interstate conference on forestry recommended that each state should reserve set minimums of permanent state forests to be managed on sustained yield principles. This move was adopted by the states, although the national target was not reached until 1966.

Queensland led the way in linking forestry with the deliberate retention of wilderness areas as national parks. Its *State Forests and National Parks Act 1906* enshrined conservation and renewal principles. It also provided for the proclamation of national parks in areas without enough marketable timber to warrant protection as forestry reserves. Queensland thus became the first state to set aside a national park where the bush would be deliberately retained in an undeveloped condition.

The Tasmanian government went even further: the *Scenery Preservation Act 1915* set up a board with responsibility for permanent reserves in regions of outstanding scenic merit. New South Wales, Victoria and South Australia all passed Acts for the reservation of national parks between 1898 and 1914, but these were intended for public recreation rather than preservation.

Like primary sectors generally, the forestry industry developed rapidly after the Second World War, spurred by industrial growth. The threat of a timber shortage during the boom of the 1950s and

way that ensured renewal and regeneration. Between 1907 and 1920 every state except Queensland followed South Australia's lead. At the same time, there was a growing demand for large areas of bush to be set aside as national parks for public recreation and enjoyment. New South Wales had been the first to respond to this call, establishing the Royal National Park in 1879 (the second national park in the world), this was followed

SUSTAINING A NATION

1960s led forest services, along with companies and individuals, to increase plantations of introduced softwoods, particularly radiata pine. But growth lagged behind increased demand and Australia continued to import timber.

During the 1960s, the management and use of forests became increasingly politicised. The demand for timber resulted in native forests being severely overcut. The demand for pulp added to the pressure on native forests; woodchips came from these, as private plantations of natives were almost extinct.

Australia began exporting woodchips to Japan, hoping to develop this industry and maintain the livelihoods of forestry workers. Large scale clearfelling of old growth areas was carried out, to keeps costs low, with even aged crops of fast growing trees replacing them. Woodchipping and clearfelling became the focus of major environmental battles, with conservationists fighting to have forests declared World Heritage areas or national parks and forestry workers fighting for a continuing, sustainable industry.

Years of dispute between the Commonwealth and the state governments produced the National Forestry Policy Statement in 1992. The total area of native forest was estimated at more than 155 million hectares in 1997, in addition to which there were 1.3 million hectares of plantations, mainly softwood.

OPPOSITE: Timber getters felling a mountain ash tree in Tasmania at around the time of Federation.

ABOVE: Felling with a mechanical saw in the 1950s. Forest harvesting changed greatly after the Second World War and from about 1950 the chainsaw replaced hand tools for felling and cross-cutting.

SUSTAINING A NATION

LEFT: A forestry operation on New South Wales' coast around 1910–20. Dairy and mixed farms were to develop in this area when the timber getters moved out.

BELOW: A sawmill in the 1920s.

OPPOSITE PAGE: Making packing boxes in the 1930s, probably using timber from eucalypts. Eucalypts supplied all Australia's requirements for strong, durable timber, including railway sleepers, poles, beams, girders, wagon scantlings, posts, wheelwright timbers, fuel and to a large measure general building and packaging requirements. The demand for timber for these purposes has declined in the past fifty years with the introduction and use of alternative materials such as concrete and steel. Timber has been largely replaced by cardboard, plastic and polystyrene for packing boxes.

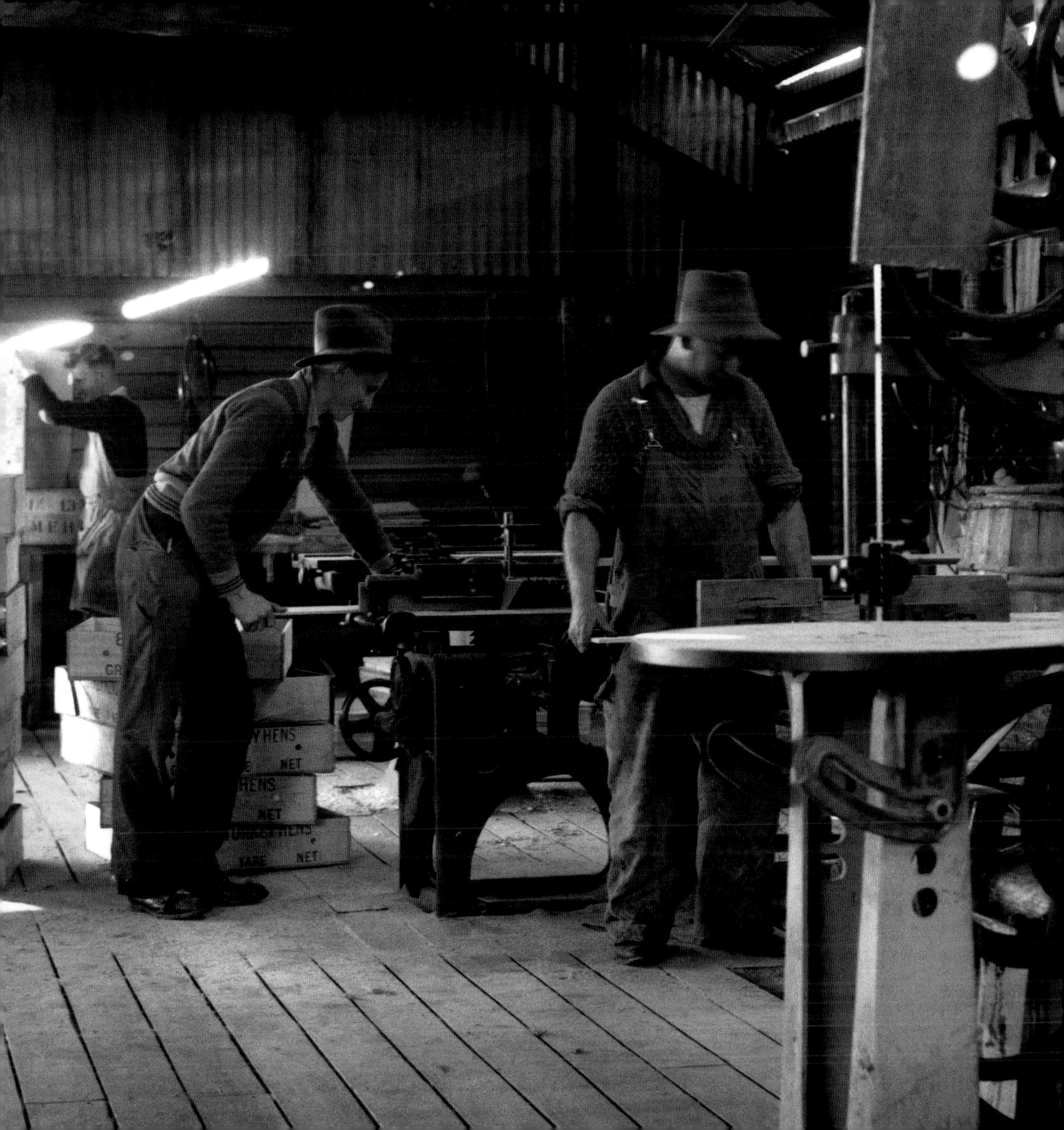
HENS
NET

ABOVE: Sawn timber being loaded in the 1920s. Shipping was the main form of transport before the extension of roads and rail.

RIGHT: Harvesting hardwood logs in a eucalypt forest. In native forests, modified earthmoving equipment is often used. Almost all plantations are harvested mechanically. Tractor-mounted hydraulic shears were first used in plantation forests from the 1970s, quickly followed by wholly mechanised logging systems.

SUSTAINING A NATION

ABOVE: Production of forest products has increased although the number of people employed in the industry has fallen as a result of mechanisation.

RIGHT: The expansion of the paper industry in the 1950s led to an increase in the demand for pulp. Certain species of eucalypts are used locally for paper making or exported as woodchips, principally to Japan.

SUSTAINING A NATION

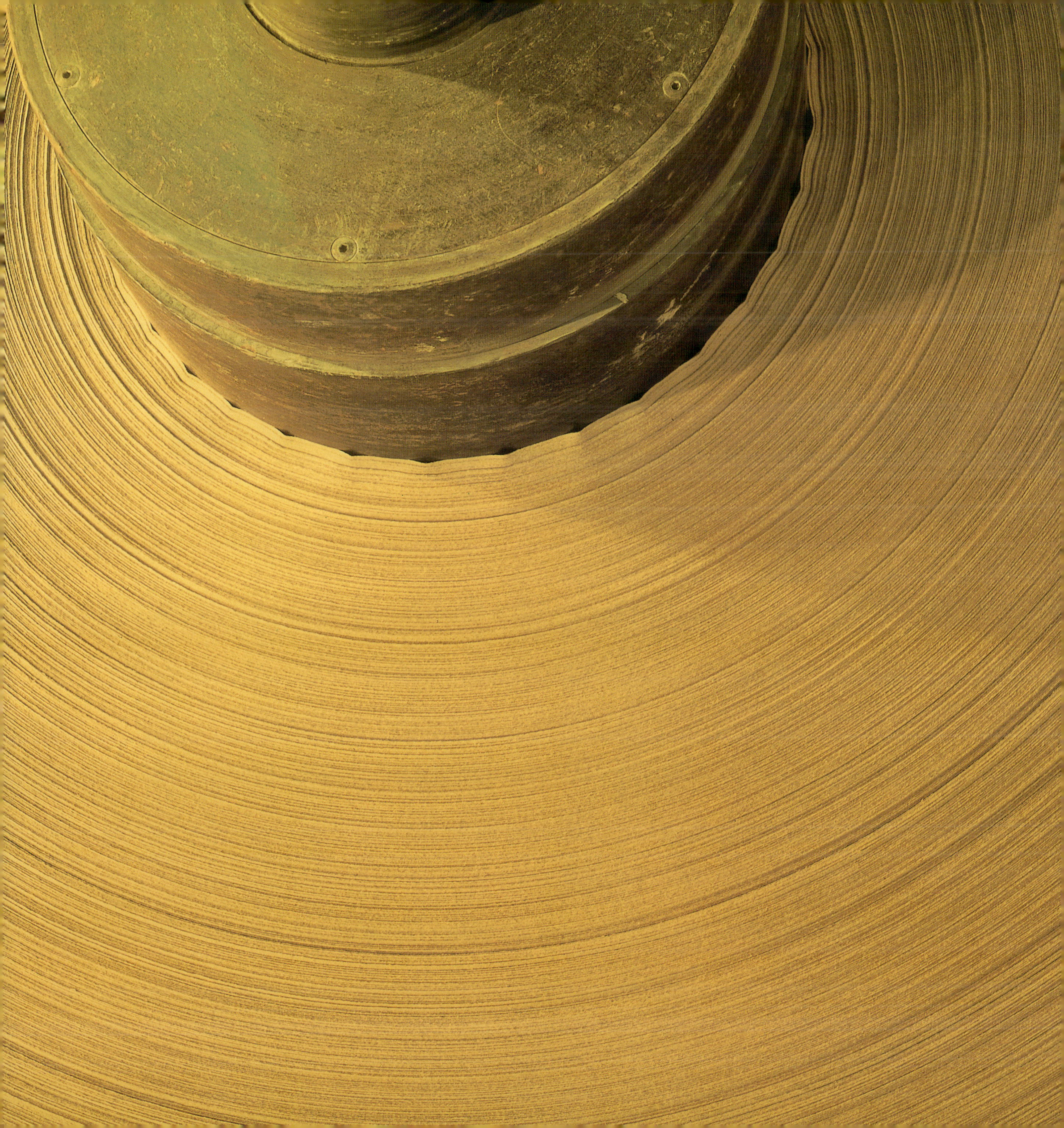

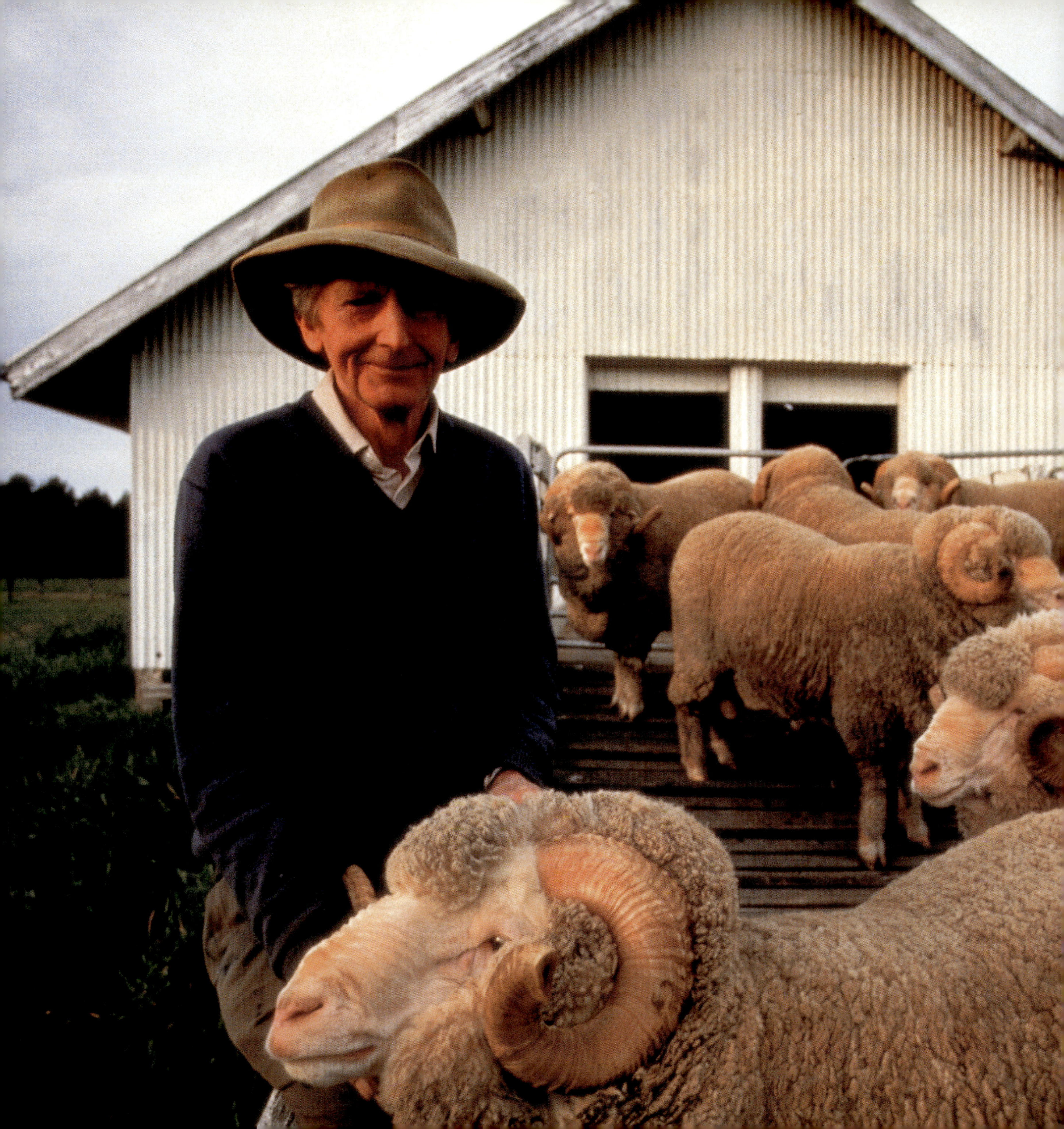

The Wealth of the Land

FIBRES

'Merino rams were a symbol of national prosperity and even of national integrity. Several generations of Australians were taught to venerate not lions or eagles or other aggressive symbols of nationalism; they were taught to venerate sheep.'

— Donald Horne, *The Lucky Country*, 1964

OF THE COUNTLESS school projects on Australia's primary industries children were asked to do in the 1950s, those on wool had the most resonance. From childhood, we knew that there was no substitute for wool. And there was good reason for this — wool contributed well over a third of our total export income until the 1960s.

Wool has had a certain symbolic importance in Australia's culture, inspiring myths and legends that are a part of our national identity. Even a brand of petrol honored the 'golden fleece'.

Until the 1960s, the wool industry enjoyed a prosperity unmatched by any other agricultural enterprise in the country. Even during the Depression, the fall in demand was not significant enough to cause a great buildup of stocks. During the Second World War we entered into an agreement with Britain in which they bought the proportion of the Australian clip not used for domestic consumption. After the war, wool prices rose and accumulated wool stocks were sold by the joint organisation with profits distributed to the partner governments — the United Kingdom, the United States, New Zealand and Australia.

The good times continued. The outbreak of the Korean War in 1950 saw a sudden demand for wool by the United States for military uniforms. During the 1951 boom, wool was selling at 'a pound for a pound'.

In 1931, Australia began promoting its wool to the rest of the world with the Empire Wool Conference in Melbourne. Six years later the International Wool Conference was formed, for the same purpose. The marketing of wool culminated in

OPPOSITE: Grazier at Goolgumbla Station, New South Wales, with his prize rams.

This chapter is proudly brought to you by Twynam Agricultural Group.

SUSTAINING A NATION — 91

RIGHT: Sheep dipping in the 1920s. External parasites such as lice, mites, ked and blowfly strike were controlled by fully immersing or 'plunging' sheep into an arsenic dip. The application of chemicals along the sheep's backline has replaced the plunge and shower dips. But blowfly strike still costs the sheep industry millions of dollars a year. While attacked by animal welfare groups as cruel, the mules operation — cutting away loose folds of skin to remove the wrinkles — remains the most effective method for preventing blowfly strike.

the launch of the Woolmark logo in Europe in 1964 — now one of the best known symbols in the world.

But by the 1970s, the halcyon days were over. With increasing competition from synthetic fibres (favored by the fashions of the decade), demand for wool began to decline. In 1970 the Australian Wool Commission established a reserve price scheme to protect wool growers from the instability that the auction system of marketing produced.

In 1973, in the face of declining demand, the Australian Wool Corporation (which had taken over the Commission's functions) adopted a fixed reserve price scheme which necessitated buying in substantial quantities of wool.

But with the escalating debt in the wool industry and a growing stockpile — and little prospect of either being reduced — the scheme was removed in 1991.

SUSTAINING A NATION

Wool has also stimulated significant research into textile technology. Australian ingenuity produced major advances, notably the 'self twist' spinning process developed in the 1960s — one of the most important advances in the textile industry in the past two hundred years.

COTTON A commercial cotton growing industry emerged in Australia in the 1960s. Attempts to establish an industry before this had faltered, with any shortlived success being linked to disrupted US production. With the uncertainty of supply from the United States during the Second World War, Britain had encouraged cotton growing in the Empire. There was a federal cotton bounty for growers, but it was only in Queensland — where cotton growing was part of the state's rural development program — that an industry had become established in any way.

With the construction of the Keepit Dam and the arrival of US cotton growers in the Namoi Valley in the early 1960s, a viable industry was established. Capital intensive irrigated cotton growing on a commercial scale spread west to the Macquarie and Darling rivers, and into Gwydir and McIntyre in the late 1970s when the Copeton Dam was completed.

OTHER FIBRES The story of Australian natural fibres does not end there. The angora wool industry — which was substantial in the late 19th and early 20th centuries — has been re-established, and the deer industry (velvet is derived from the antlers) has also made an impact since the 1970s.

ABOVE: Merino rams. Pastoral landscapes such as this came to be considered 'typically Australian'. But, while Australia might have ridden to prosperity on the sheep's back in the 1950s, it came at a price. Unlike native fauna with pads, imported livestock with cloven hooves damaged Australia's fragile natural environment.

SUSTAINING A NATION

RIGHT: Shearers using machine shears, 1908. The introduction of machine shears from 1888 was initially opposed by the unionised shearers. A few conservative pastoralists also opposed the shearing machine. An electric shearing machine was invented in 1906. By the outbreak of the First World War, steam driven shearing machines had been installed in most large sheds. The shearer — along with other skilled rural workers — was considered to embody important Australian characteristics such as mateship and egalitarianism. Shearers played a major role in the development of organised labor in the 1880s and 1890s.

OPPOSITE: Apart from changes brought about by machine shearing, little has changed in the shearing process in the past hundred years, other than the admission of women as rouseabouts to what was once an exclusively male domain.

SUSTAINING A NATION

SUSTAINING A NATION

TOP RIGHT: Children of Aboriginal shearers astride a sheep in front of the shearing shed at 'Barokaville', Walgett, around 1936. The contribution of indigenous Australians to the pastoral industry throughout Australia has included their many roles as shearers, stockmen, drovers, fencers, cooks and providing household help.

RIGHT: A knitting competition in the 1950s. As Alexandra Joel in her history of Australian fashion has noted: 'During the (Second World) War women had become used to knitting socks, gloves and balaclavas for the troops ... It remained a craze, with home knitters reluctant to down needles once the fighting was over.' The 'baby boom' in the postwar years also kept those needles clicking.

96 — SUSTAINING A NATION

ABOVE: While this is probably a promotional image, such an event would not have been unusual. A popular myth from the wool boom years was that wealthy graziers transported their sheep in the back of their newly acquired Bentleys!

RIGHT: Long haired dogs were not suited to work in Australia's heat and distances, which led to the development of the kelpie which, along with the border collie, is one of the two main sheepdog breeds in Australia.

SUSTAINING A NATION 97

ABOVE LEFT: A cotton plot in the Dawson Valley, Queensland, in the early 1920s. Until the 1960s, cotton growing was almost solely confined to small scale production in Queensland.

LEFT: A modern cotton gin removing the impurities before the cotton can be dispatched to the spinning mills.

RIGHT: Mechanised cotton harvesting. The absence of cheap labor in Australia militated against the development of a viable cotton industry until the 1960s when US growers introduced capital intensive irrigated cotton growing to the Namoi Valley in New South Wales.

SUSTAINING A NATION

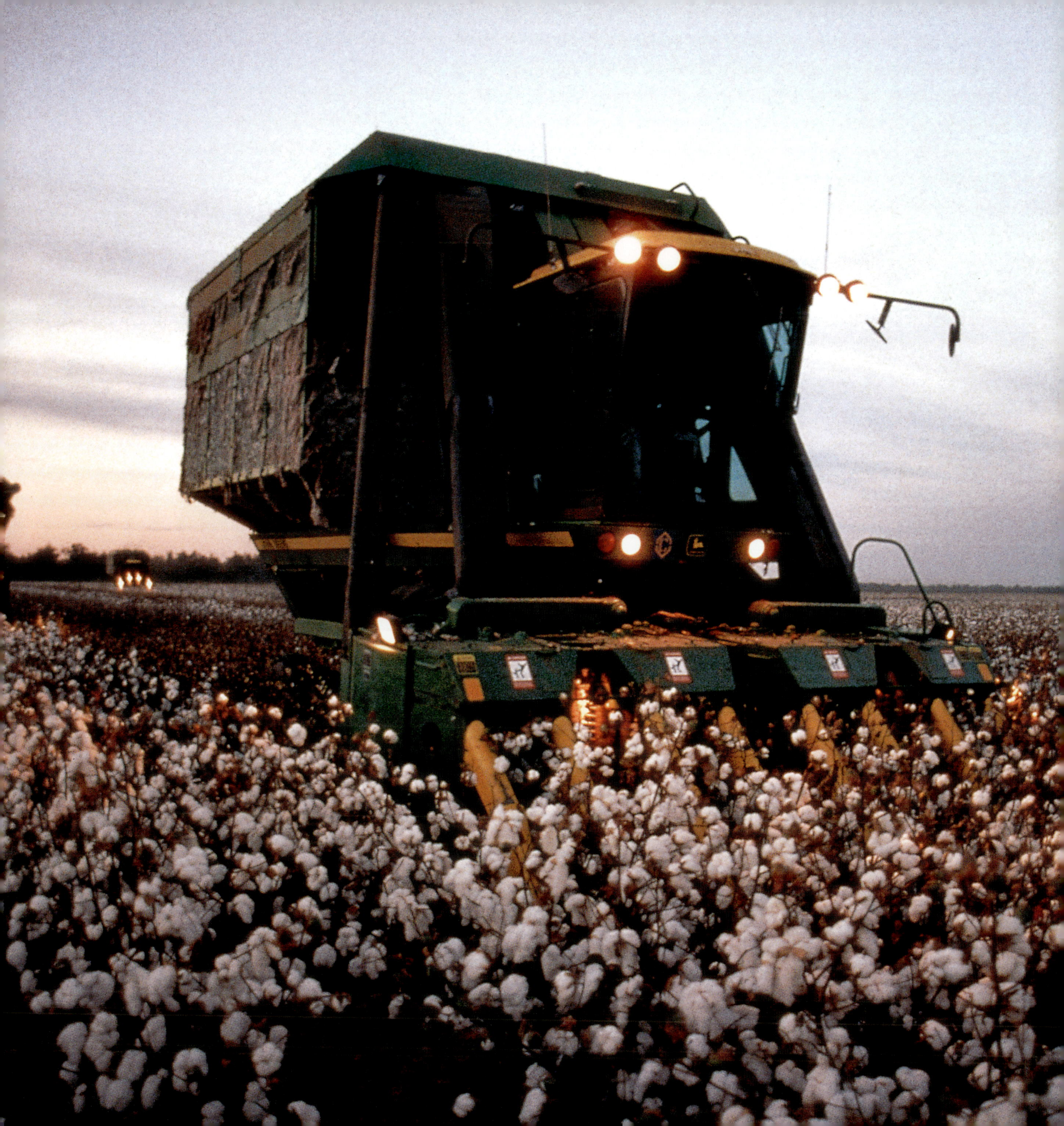

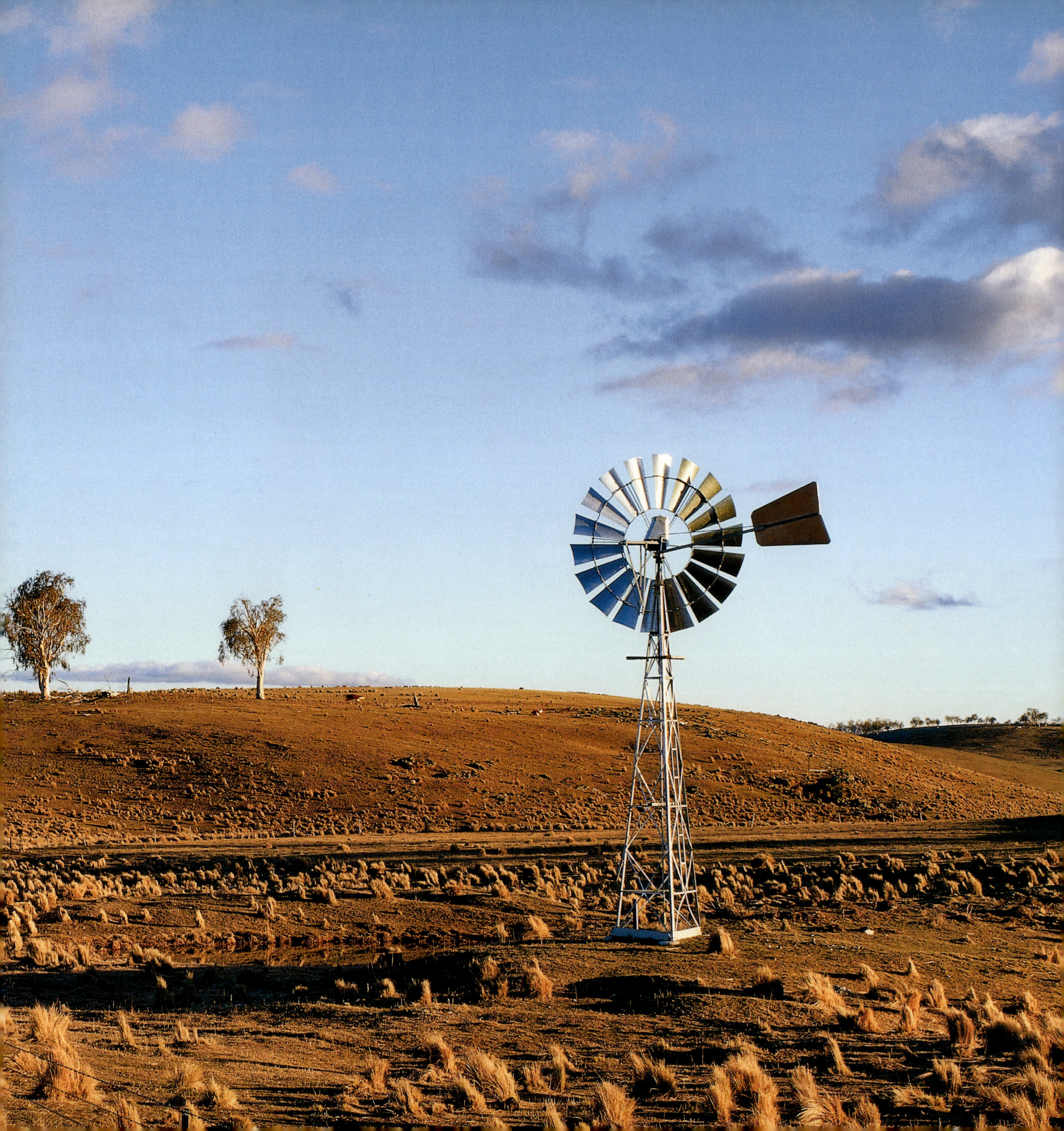

PART TWO

CHALLENGES & OPPORTUNITIES

'There is an important need to create profitable enterprises that are attractive to younger people and help stimulate the local economy. However, there is an equal need to thoroughly research new ventures in the current economic climate for there is no room for mistakes.'

— Roger Epps, *Prospects and Policies for Rural Australia*, 1993

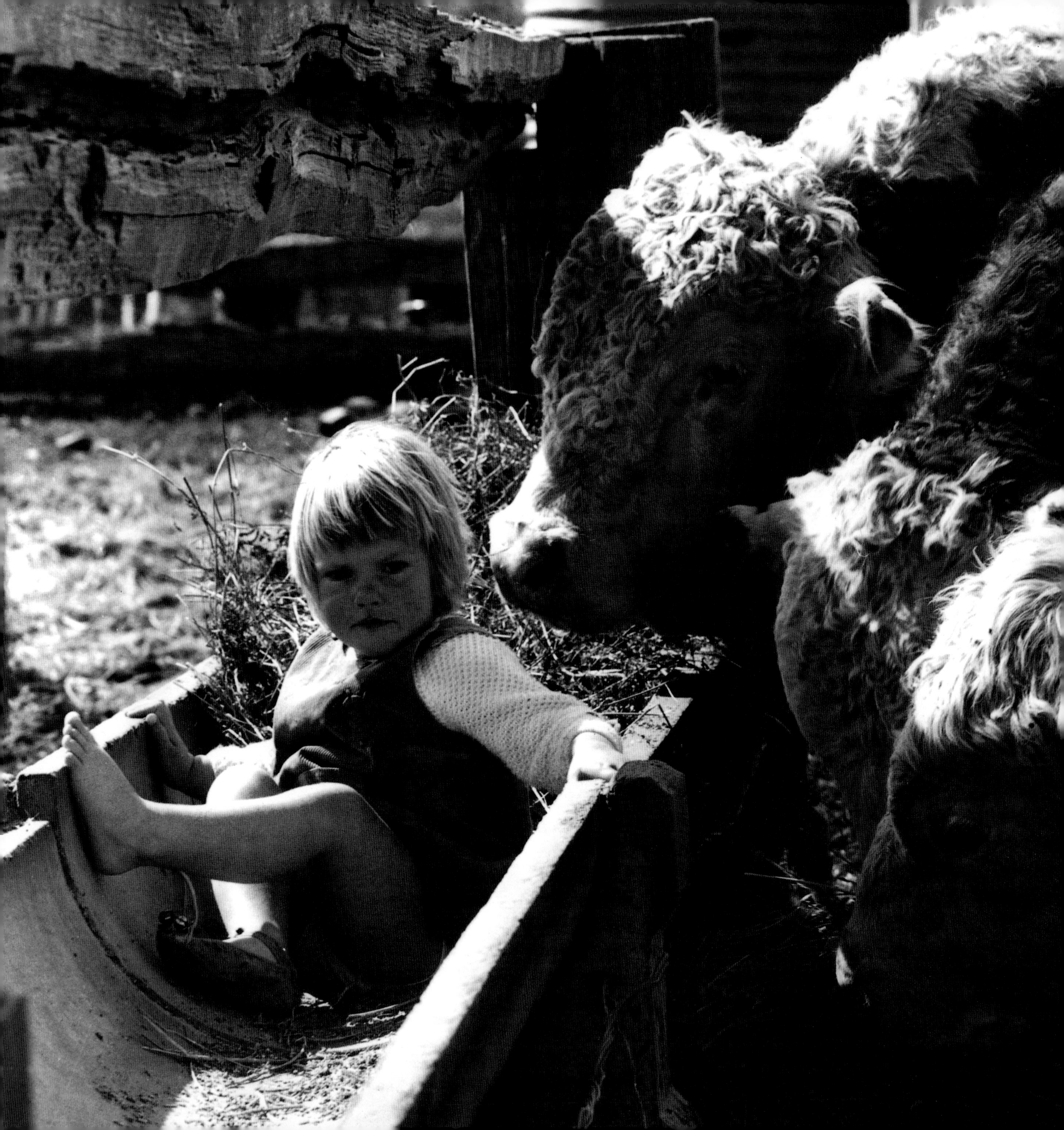

Challenges and Opportunities

LIFE ON THE LAND

'The traditional public image of country towns as primarily sleepy and relatively unchanging service centres for primary producers is misleading. In practice, the picture is much more complex and dynamic. Some people face a bright future. For others, their livelihood is at stake as they confront loss of jobs and services, decaying environments and declining property values.'

— Tony Sorensen, *Prospects and Policies for Rural Australia*, 1993

THE IDENTITY OF ANY BUSH COMMUNITY is its local town. Established as service centres for surrounding agricultural communities, the country town in the early 20th century had a police station, a bank, usually two churches and a post office. Larger towns could also boast a railway station, a school, a local newspaper and a mechanics institute. The main street, with its wide footpaths and post-supported verandahs, was where news and gossip were exchanged and business discussed — the weather, prices and prospects for the season.

The saleyard, sports oval, showground and town hall (and perhaps a race track and bowling club) were the local amenities along with several pubs, a picture theatre, a (Greek) café (or fish and chip shop) and the Returned and Services League club. The 20th century town also had a war memorial and sometimes a cenotaph or clock tower. After the Second World War in the more prosperous 1950s, these local facilities were extended to include hospitals, libraries and swimming pools.

The most important annual event was the agricultural show. Here the latest machinery or farming practices were demonstrated and displayed; livestock and produce competed for prizes, benchmarking best practice and excellence. Agricultural shows in Australia had their origins among forward-looking farmers who formed agricultural and pastoral societies to improve farming and animal husbandry, both in methods and machinery. By the late 19th century the agricultural society movement had begun to spread.

OPPOSITE: Farming was not just a way of earning a living, it was a way of life as promotional images such as this show. State government advertising after Federation often promoted Australia as 'a land of freedom and opportunity'. However, the life of children on the land was not always carefree. Child labor was commonplace on small, struggling farms.

This chapter is proudly brought to you by National Farmers' Federation *and* Rural Press Limited.

SUSTAINING A NATION

ABOVE: A meeting of the Country Women's Association in the 1950s. The CWA, as it is generally known, was established in New South Wales during 1922 largely due to the efforts of Grace Emily Munro. The local CWA branch is still a central institution in rural communities.

The show also provided a valuable bonding function for rural communities. It was an opportunity to express local pride and identity. As the crowds at shows increased to include nonfarming groups, amusements began to appear. Despite much resistance from the agricultural societies, sideshow alley (which had been transplanted from the fairgrounds of Britain) began to appear, becoming a permanent fixture of shows by the 1920s. Along with the vaudeville circuit, the boxing tent and the circus, sideshow alley became an important form of travelling entertainment for country areas.

Nearly every town of any size had (and many still do) a branch of the Country Women's Association. Founded in New South Wales in 1922 and faithful to its motto 'honour to God, loyalty to the throne, service to the country through country women, for country women, by country women', the CWA concerned itself primarily with the welfare of women and children, with activities ranging from providing baby health care centres and holiday homes at the beach to handicraft instruction and music and drama groups. As a national lobby group it has secured significant improvements to government services in rural areas over the years, particularly relating to health care and education.

Falling membership of the CWA today and its waning political influence is symptomatic of the changing rural landscape. Many bush communities are now struggling to survive under the changes wrought by farm rationalisation and globalisation.

It is perhaps the role of women that symbolises most the changing face of agriculture. Women have always played a central role as bookkeepers, farmer's wife and committed nurturer in bush communities. Today women — with their broad understanding of issues facing the rural sector — are bringing a diversity of ideas and skills to the challenges confronting rural Australia. They are now clearly visible in the operations of farm business.

Mechanisation and economic necessity have led to the decline of the once large rural worker population (farm laborers, cane cutters, stockmen and sleeper cutters to name but a few) and farm consolidation and corporatisation have meant a further contraction in the population of bush communities. Inefficient or small scale operations have given way to fewer but larger enterprises.

And like the farms, small country towns have also been a casualty of consolidation, a process that began long ago in the 1920s with the increasing use of the motor car.

These days, the internet is a useful tool on the farm. And accessing 'virtual markets' is becoming as important as weather forecasts. Farmers have developed new skills in a wide range of areas including financial and risk management, natural resources management, information access and understanding and interpreting market signals.

Productivity growth has arisen largely through better production methods, new technologies and farm amalgamations. The resilience and adaptability of the Australian farmer can be seen in areas such as the southern Riverina where successful rice growing has transformed the region. Other examples include the success of oilseed crops, cotton and wine. And the creation of value adding industries in rural areas — such as food processing — are also creating new opportunities for local populations.

ABOVE: Judging a CWA sponge cake competition in the 1950s. One of the CWA's many roles was guiding the 'woman on the land' in the domestic sphere. Keenly aware of the challenges of domestic life for the country housewife, CWA branches produced cookery books that often became best sellers. The CWA was also called on to cater for local events.

SUSTAINING A NATION

RIGHT: Display of machinery at a field day in the 1940s. Field days provided a chance to exchange information about the latest technology and farm practices and were often held at government experimental farms. They were both educational and social events, although distance, time and sometimes prejudice, tended to keep many farmers away.

OPPOSITE: The shearing competition was a popular and hotly contested event at agricultural shows and the status of ringer (fastest shearer) was highly esteemed. Competitions such as this provided the opportunity for rural workers to display their skill publicly. The show was also a useful vehicle for promoting products and practices. On this occasion, the shearers are doing just that — using the popular Lister shearing machine.

But a divide has emerged between those who have embraced the challenges and those who pine for the good old days. As one commentator observed of rural communities in 1993, 'farm households typically comprise a couple in their fifties who subsist on the questionable benefits of a way of life during hard times'. This divide in Australian agriculture can be further broken down into those who see farming as a 'way of life', and those that believe that farming is — like any other 'way of life' — the responsibility of the individual to work at and succeed.

Agriculture in the 21st century is a business geared to meet the demands of ever changing markets rather than a way of life, as long epitomised by the family farm.

Australian farmers, graziers and growers are meeting the challenges and maintaining their position as running one of the most successful farm enterprises in the world.

SUSTAINING A NATION

LEFT: A prize winning Clydesdale at a country show. The draught horse was essential for farm transport until after the Second World War when the tractor became the principal 'tractive unit' on the land.

OPPOSITE: Spruiking to the crowd in sideshow alley in the 1950s. The travelling sideshow alley was a fixture at agricultural shows by the 1920s, and also provided entertainment at country events like picnic race meetings. The general suspicion and distrust of outsiders by bush communities was even more marked in the case of itinerant show people and the saying, 'lock up your daughter and your chooks, the showies are coming', was only half in jest.

RIGHT: A rodeo in the 1950s. 'Bushmen's carnival' was the name originally given to what, from the Second World War, became known as rodeos. American influences have been many and varied, particularly in remote rural areas.

108 — SUSTAINING A NATION

SUSTAINING A NATION

RIGHT: This heroic image was part of a propaganda campaign in the Second World War to attract women to work on the land thus releasing men for war service. Here a member of the Women's Land Army (formed in 1942) shows her skills in the driver's seat of a Farmall tractor. But such smiles could be misleading, as many city women were completely unprepared for the hardships of life on the land.

LEFT: Elsie Clapham about to take a dip in her woollen cossie on a hot February day in 1937 at Hepburn Springs, Victoria. Creeks, rivers, dams and irrigation channels were all popular swimming holes during hot bush summers. But they were often treacherous and there were numerous drownings. Public swimming pools did not become common in country towns until the 1950s.

RIGHT: The role of women symbolises the changing face of agriculture. They are clearly visible today in the operations of farm business, bringing a diversity of ideas and skills to the challenges facing rural Australia.

PAGES 112—13: Prospective buyers inspecting cattle from the safety of the saleyard ramps.

SUSTAINING A NATION

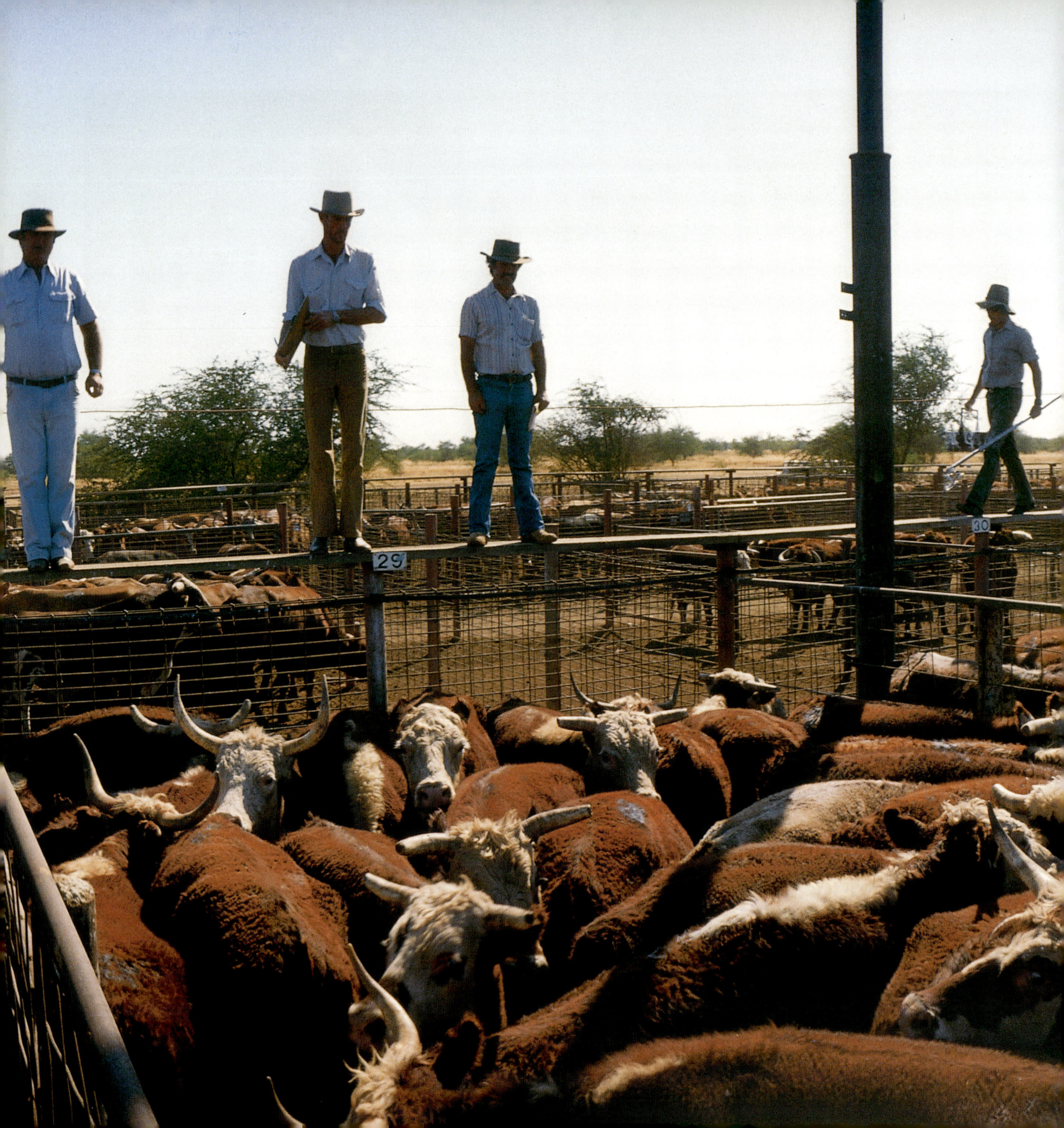
29
30

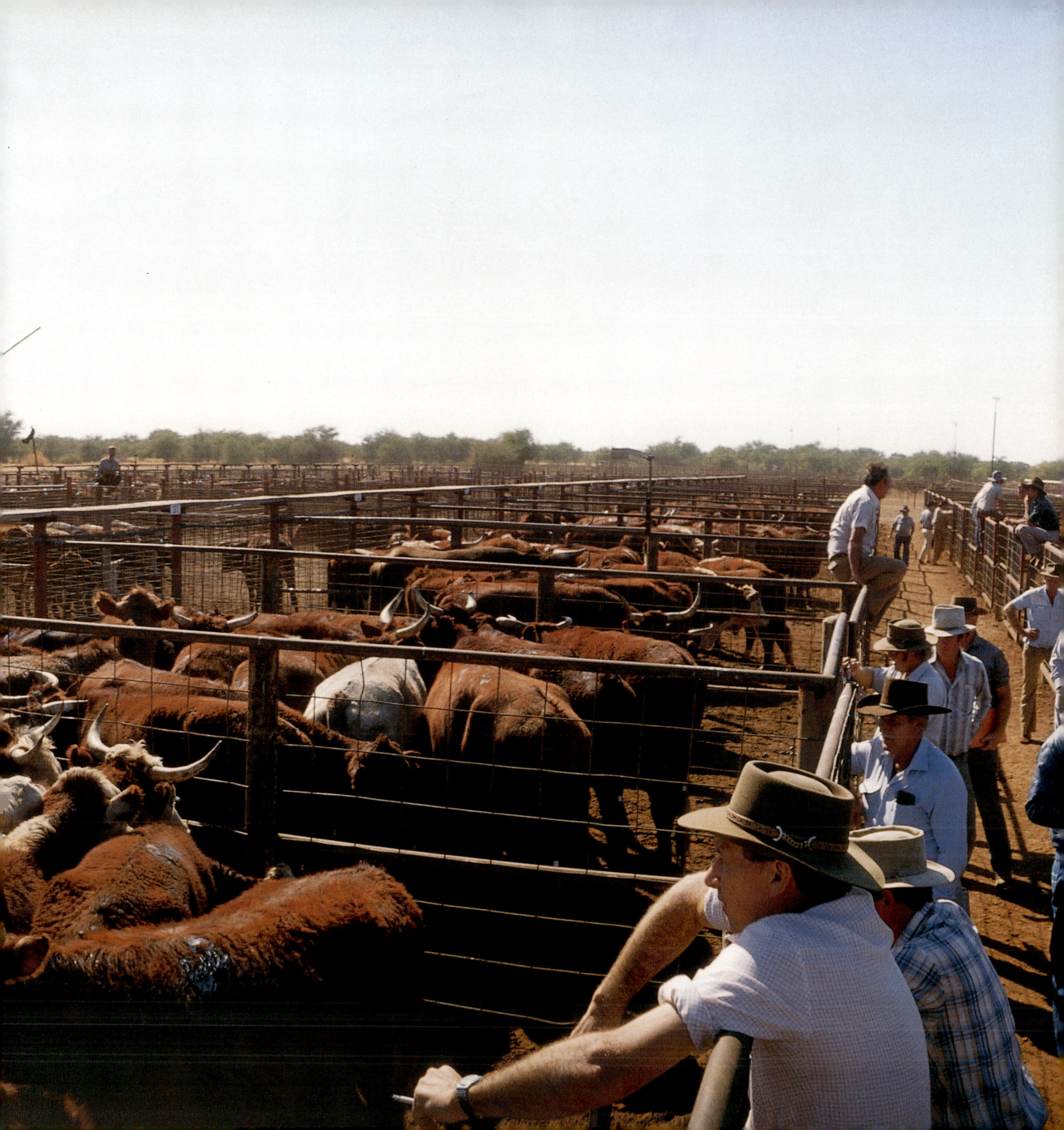

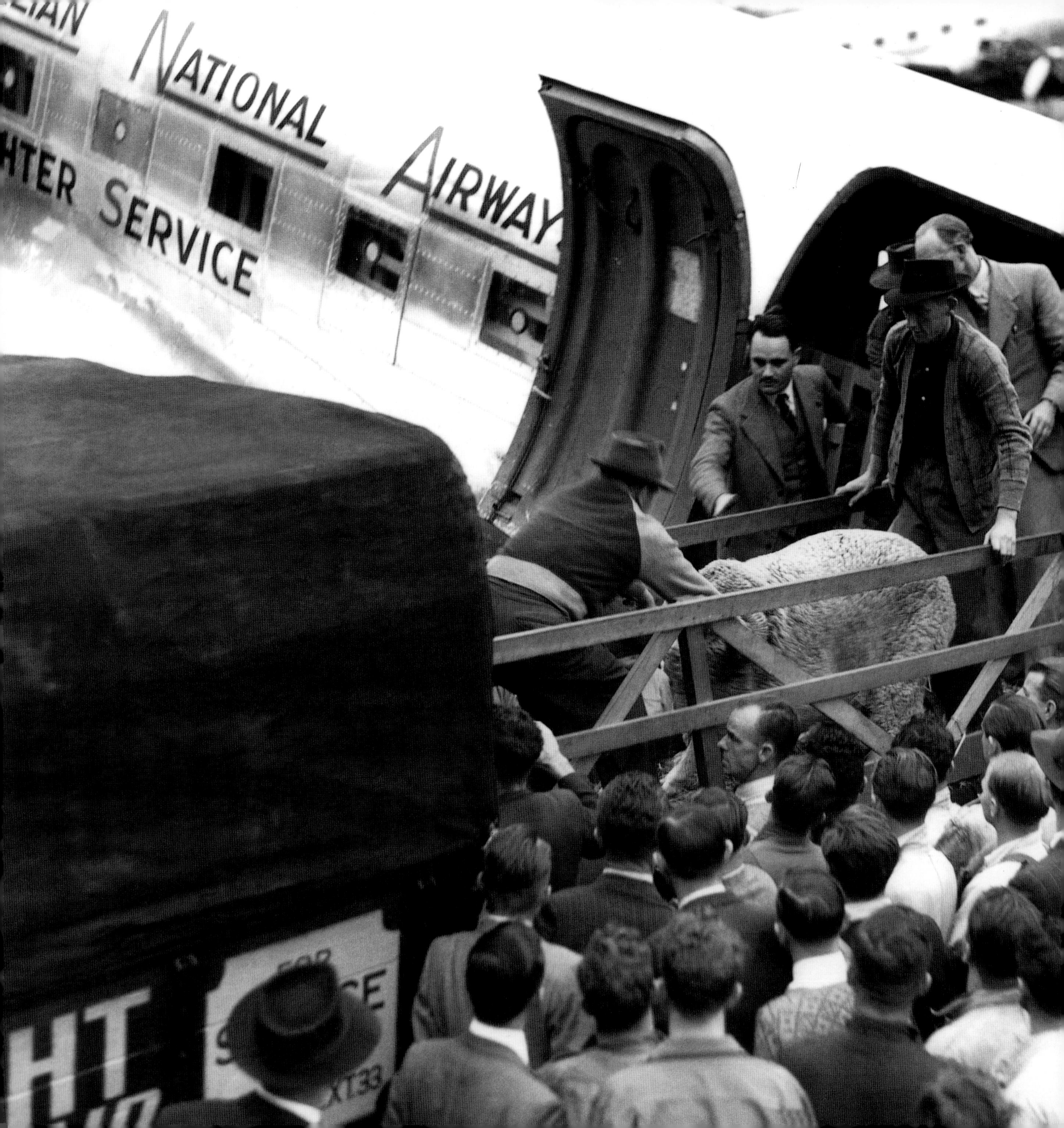
NATIONAL AIRWAY
SERVICE

Challenges and Opportunities

GETTING GOODS TO MARKET

'Packhorses, except for the solitary traveller, were replaced by the bullock-wagon. It was slow travelling, but the great-hearted beasts could be relied on to surmount all obstacles and the bullocky's verbal encouragement was proverbially lurid.'

— Marjorie Barnard, *A History of Australia*, 1962

IN THE NEW WORLD of free trade and globalisation, Australian agricultural producers have learnt to 'grow for their market', not 'market what they grow', which is why they are so successful. Agriculture is now a business that takes producers beyond the farm gate and into the market — distribution has assumed an importance equal to the production process itself.

For much of the 19th century the distances involved and the available transport restricted the opportunities for developing and expanding markets for agricultural produce. Bullock trains hauled bags of harvested crops and wool bales along rough roads from the farm to railway points and wool stores. Drovers overlanded large mobs of cattle and sheep along inland stock routes on journeys taking months to pastures in the south for fattening or to the meatworks. And the only markets for perishable products were local populations.

Railways, refrigerated transport and cold storage — freezing in the 1880s and chilling in the early 1930s — created new domestic and export markets, for perishables — meat, dairy and fruit particularly. After Federation, large scale public investment extended railways, built roads and improved stock routes by sinking bores.

The greatest impact of changes in transport has been in remote pastoral areas. The use of motorised road transport in northern Australia (road trains) made possible by the construction of roads during the Second World War would see the drover completely disappear by the late 20th

OPPOSITE: A sheep being coaxed off an Australian National Airways freighter in the 1950s. Merino rams were so valuable at the height of the wool boom in the early 1950s as to justify air freight.

This chapter is proudly brought to you by FreightCorp.

ABOVE: The milkman making a delivery to the Blackfriar's Infant School in the inner city suburb of Chippendale, Sydney. Distribution districts (or 'milk runs' as they came to be known) were regulated in some states to wipe out route duplication by competing vendors which were a source of price fluctuation. These arrangements were continued among milk vendors. The milkman making his daily deliveries has now all but been replaced by the supermarket and corner store.

century. And the negotiation of the British meat agreement in the early 1950s led to 'Air Beef' — a system set up to fly carcasses direct from cattle stations to meatworks.

Handling systems were also an important factor in the circulation of commodities. Along with the introduction of the bulk handling of wheat in 1920–21 in New South Wales, a state financed and controlled silo system came into operation. By the 1950s all other states had followed. The sugar industry was also converted to bulk handling by 1964. Similarly, the development of bulk and container shipping in the postwar period meant that shipping continues to be a very important mode of transport for many agricultural products including grain.

Government intervention to help the 'man on the land' has been in the nature of farming in Australia. Throughout the 20th century, governments intervened in the operation of markets. The First World War opened the way for unprecedented levels of regulation with the introduction of emergency controls to ensure efficient handling of produce to supply the armed forces. The first compulsory wheat pool, for example, was set up in New South Wales in 1914, and from 1915 until 1922 all wheat produced in Australia was compulsorily pooled. Such wartime measures laid the basis for future regulatory mechanisms.

Wartime 'marketing' also resulted in rural industries being tied to government purchasing schemes. Demand for agricultural produce in the postwar period created optimism about these markets. High butter prices in the 1919–20 season by the Ministry of Food, for example, greatly inflated land values for dairying, ultimately compounding the problems of overcapitalisation that were to be dearly paid for on newer dairy farms throughout the 1920s.

Many government interventions and schemes were to prove inefficient. Poor administration and lack of technical advice and expertise were dearly paid for by many on the land.

From the 1960s, changing trade relationships created a need to develop new markets and diversify production. The dismantling of protective trade barriers and price support schemes has meant that Australian agriculture has been directly exposed to global markets. The export market was no longer a dumping ground for surplus production, but a highly specialised one requiring producers to meet specialised demands and produce value added goods.

More recently, technology has changed the imperatives of physically getting certain goods to market. The advent of 'sale by sample' in wool where buyers inspect objectively tested wool samples rather than visually appraising whole bales at selling centres has revolutionised selling and marketing. But the latest innovation — 'sale by description' — has created 'virtual markets' for a number of goods, including meat and wool, through the use of communications technology, particularly the internet.

Getting goods to market today means getting them there fast, fresh and in perfect condition. Australian producers are very aware that to remain competitive in the global marketplace they have to be at the cutting edge of new packaging and transport technologies and up to date with global trends and market opportunities.

ABOVE: Merino rams being loaded into railway wagons at Sydney's Darling Harbour goods yard in 1948 by the Graziers' Co-Operative Society after competing in the Royal Agricultural Show.

SUSTAINING A NATION

TOP LEFT: Field workers top and tamp wheat bags prior to sewing them. Each bag weighed 3 bushels, or approximately 180 pounds (82 kilograms).

LEFT: Loading wool bales onto the back of a truck outside the shearing shed in the 1930s. Using a 'block and tackle' system of pullies enabled the men to lift the 300 pound (136 kilogram) wool bales.

RIGHT: Following the introduction of bulk handling in the 1950s, bagged wheat was carted to one of the depots located throughout the grain growing areas. The sewn bags of wheat shown here would have been tipped into a grate and augered into the silo by a grain elevator. Open bags, however, were the norm until farmers incorporated bulk handling facilities (such as bulk bins) in the harvesting process and the wheat bag disappeared altogether.

118 — SUSTAINING A NATION

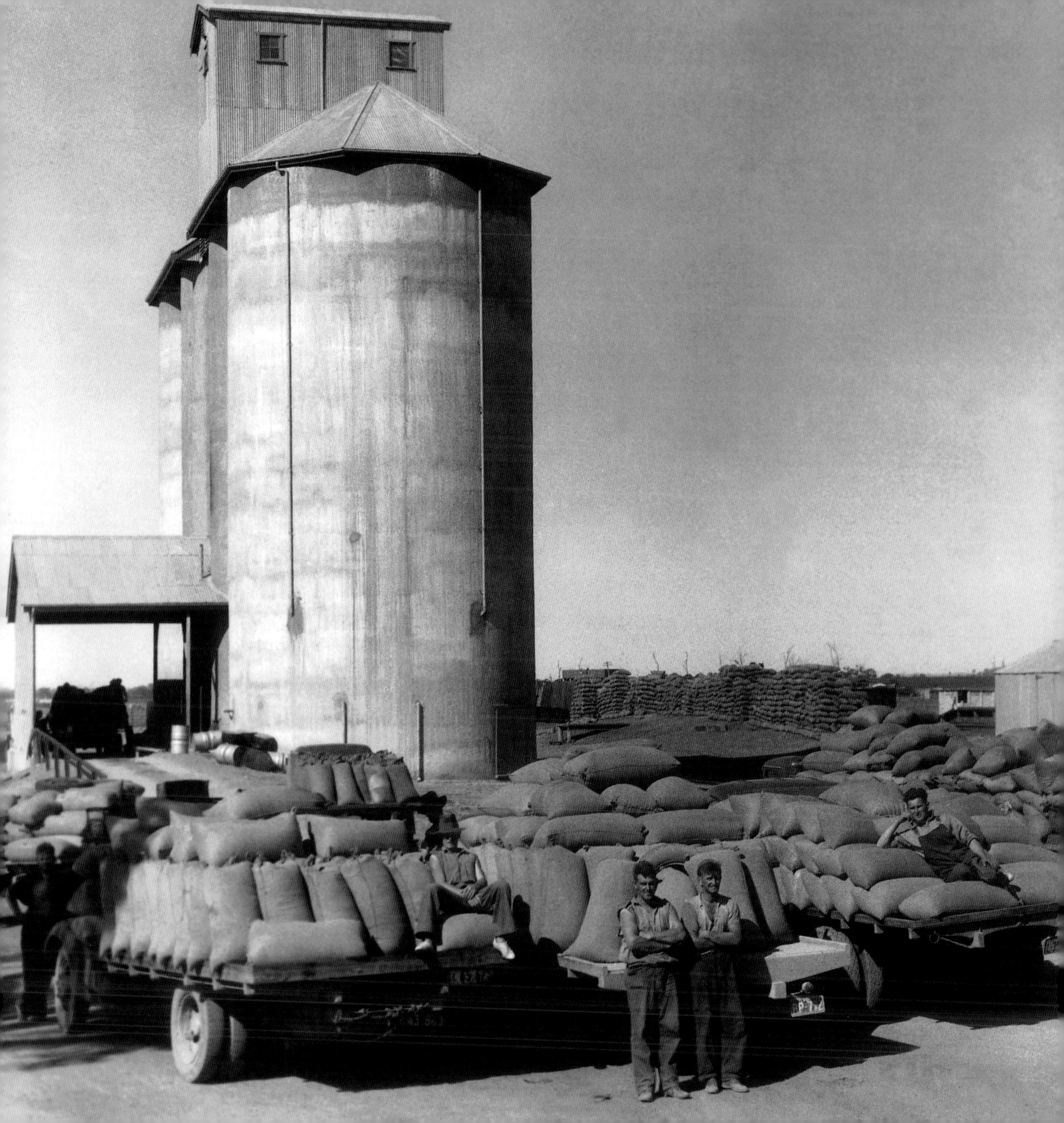

ABOVE: Loading sheep for export to Saudi Arabia 'on the hoof'. A major expansion of the live export trade occurred from the 1960s arising from the Middle East demand for 'hot' (freshly killed) meat.

RIGHT AND OPPOSITE: Shipping has remained an important form of transport for grain. These images show loading wheat into a bulk shipping facility for export today (right) and wool being winched on board ship (opposite) in the 1920s.

120 — SUSTAINING A NATION

SUSTAINING A NATION

LEFT: Weighing bagged wheat prior to transport. Before the advent of bulk handling, wheat was transported in bags that were 'topped and tamped', sown up and weighed in the field. This photograph was probably taken in the late 1920s.

BELOW: A promotional photograph from the abattoir at Blayney, New South Wales, in the 1950s. Refrigerated transport meant that producers of meat and other perisherable products were no longer limited to local markets. Refrigeration also helped the development of fruit growing in Tasmania (apples) and in the irrigated districts of New South Wales and Victoria. The advent of chilling in the mid-1930s increased the export potential of meat and other perishables even further.

SUSTAINING A NATION

ABOVE: Road trains for transporting stock appeared after the Second World War and made use of roads that had been constructed during the war throughout Australia's north. By the late 20th century, the road train had replaced the need to overland large mobs of cattle and sheep along stock routes. Combined with rail transport, the road train has enabled stock to be turned off at an earlier age and has reduced weight deterioration and loss of livestock in transit.

SUSTAINING A NATION

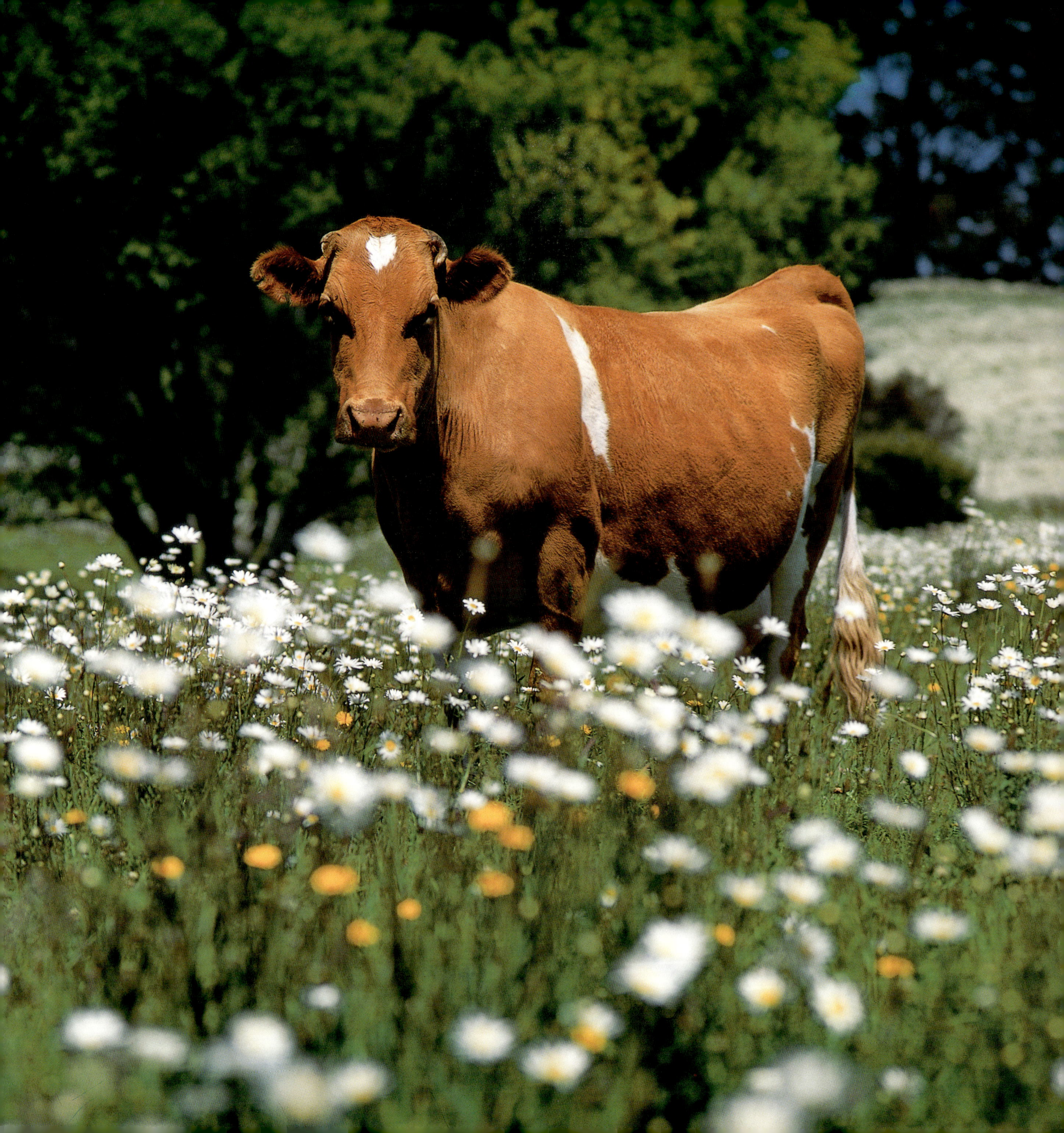

Challenges and Opportunities

SUSTAINABLE AGRICULTURE

'We toiled and toiled clearing those four acres, where the haystacks are now standing, till every tree and sapling that grew there was down. Dad was never tired of calculating and telling us how much the crop would fetch if the ground could only be got ready in time to put it in; so we laboured the harder.'

— Steele Rudd, *On Our Selection*, 1899

FARMING has always been conducted under difficult conditions in Australia, the world's driest continent apart from Antarctica. But we have virtually been pioneers of dryland farming, and we are now able to export our 'know-how' to areas with similar environmental constraints such as the Middle East.

Lord Sydney and the British powers that be had great expectations of Australia, expecting it to produce wheat, fruit, flax, tobacco, cotton with 'tropical abandon'. Thus the first white settlers arrived in Australia with high hopes of recreating a rural landscape just like the one they had left behind. Areas were settled in good seasons on the assumption that 'the rain follows the plough'. But in the attempts to expand production, areas of unreliable and low rainfall were cultivated or overstocked and environmental damage was done.

Certainly by 1901 the implications of soil erosion were acknowledged when a royal commission investigating the economic depression in New South Wales' western district identified soil erosion as a principal factor, and enumerated some of its causes: overstocking, wholesale destruction of vegetation, rabbits and drought.

In 1933 a government committee was set up to monitor the problem in New South Wales and in 1938 a soil conservation service was set up to survey its extent. Similar legislation was enacted in Victoria (1940) and Western Australia (1945), while other states conducted soil conservation schemes through their departments of agriculture.

Modern agricultural science emerged in the 1920s along with the establishment of the CSIRO (1926), largely in response to the problems besetting the industry. Major breakthroughs included the use

OPPOSITE: The serious challenges of managing the environment for sustainable agriculture and being a profitable producer constantly face Australia's farmers.

This chapter is proudly brought to you by NSW Agriculture, Victorian Department of Natural Resources & Environment — Agri-business Group, Victorian Department of State & Regional Development — Food Victoria *and* Westpac Banking Corporation.

SUSTAINING A NATION

RIGHT: Treating a pasture infested with skeleton weed in the 1940s. Over 1300 plants have been introduced during the past two hundred years, including all the sown pastures and some of the worst weeds. While some were introduced deliberately, others arrived by accident. Skeleton weed emerged as a pest with the expansion of wheat growing around the time of Federation.

OPPOSITE AND INSET: Land infested with prickly pear near Inglewood, Queensland, 1927 (inset). Declared a noxious weed in 1900, nearly 25 million hectares of agricultural land in Queensland and northern New South Wales were infested by 1925. The photograph opposite shows *Cactoblastis cactorum* eggs being 'quilled' for dispatch to landowners in Cinchilla, Queensland, in October 1929. The larvae of this South American moth were released in 1926, bringing prickly pear under control by the mid-1930s.

of subterranean clover and superphosphate to improve pastures and soil fertility, the identification of trace element deficiencies (which led to the opening up of new land to grazing, notably the Ninety Mile Plain) and improved breeding.

The control of livestock diseases has been crucial to sustaining the meat industry. Quarantine restrictions (and Australia's isolation) have always been an important control measure. But veterinary science was also to play an important role in the 20th century. The development of vaccines during the interwar period for the control of clostridium diseases in sheep such as black leg and black disease was revolutionary. In the case of cattle, major breakthroughs were made in the postwar years with the eradication of pleuropneumonia (which was endemic in northern Australia and spread to the south east when cattle were shifted down for fattening up), brucellosis (or 'contagious abortion') and bovine tuberculosis.

Australia in its virgin state had no pests of European agricultural crops and livestock. Many of the plants and animals that were introduced by settlers to remind themselves of 'home' soon made their presence felt as pests. The prickly pear — planted as a cheap and ornamental hedge as early as 1839 — was a noxious weed by the 1880s as agricultural land became infested.

The rabbit — introduced by a Victorian squatter in 1859 for hunting — was to acclimatise so well that by the 1880s it had become a serious threat to rural production. It destroyed pastures, ringbarked trees and its burrowing added greatly to problems of soil erosion. Neither a fence dug deep into the ground nor poison baits controlled their spread and their numbers reached plague proportions by the 1920s. It was not until the release of the *myxomatosis* virus in the early 1950s that the plague was brought under some control. But the virus lost its impact over time as the rabbit population developed immunity. The latest weapon in controlling the rabbit, *calicivirus*, was introduced in the mid-1990s and was particularly effective in the drier areas of the continent.

Other species were introduced as control measures against pests. The giant South American cane toad, for example, was imported to Queensland in 1935 to control the greyback

SUSTAINING A NATION

ABOVE: Spraying an orchard with pesticides around the 1950s. The use of pesticides increased after both world wars, particularly the Second World War, given advances in industrial chemistry during that time. Combating harmful infestations of pests and diseases ranked alongside irrigation in the further expansion of horticulture.

beetle. While it did help control the beetle borer and other minor pests it, like other introduced species with no natural predators here, it has spread from the cane growing areas to become a substantial pest.

The expansion of Australian agriculture has been inextricably linked with water conservation. Irrigation, often became the means of making small scale farming possible and of increasing agricultural production. But rising groundwater levels have drawn naturally occurring salt from the deeper levels of the soil to the surface causing land salinity. And disturbance of the natural 'runoff' has in turn increased water salinity. In fact, salinity was one of the earliest and most visible effects that prompted the development of better management practices.

It is now apparent that irrigation schemes have had a detrimental effect on the major river systems, especially the Murray–Darling by contributing to increased salinity, blue-green algae and riverbed erosion.

It was also not well understood that the natural flow of the river system in Australia provides annual or periodic flushing and that this is interrupted by damming.

Australians are responding to challenges of taking greater care of the country's fragile soil and water resources and rehabilitating the land and waterways with the implementation of programs such as Landcare which aims to plant one billion trees and the development of concepts such as 'environmentally accredited' properties. Most states have also introduced vegetation clearance legislation requiring landowners to seek government approval for any further clearing of native vegetation.

Individual farmers are also playing a part — planting trees, creating shelter belts, retaining native vegetation, adopting drip and trickle irrigation, practising minimum tillage and reducing the use of chemicals — in the knowledge that the continued viability of agriculture depends on the adoption of sustainable practices.

ABOVE: An artesian bore plant. Severe drought prompted a new means of conserving water. Private and public bodies pumped artesian waters to provide 'oases in the desert' on the world's driest continent. Better management of water resources remains a serious challenge for Australia.

RIGHT: Poor farming practices and clearing combined to cause this example of soil and gully erosion. Planting trees and contour ploughing will help prevent such land degradation in future.

LEFT: Land salinity, the end result of dryland farming and grazing where land has been cleared of its native vegetation and replaced with agricultural systems that use less of the available rainfall. Rising groundwater levels drew stored salt to the surface. The naturally occurring salt in Australia's topsoil was not a problem while deep rooted native vegetation was drawing most of its moisture from the deeper levels of the soil. Remedial works — the most obvious being planting trees — has begun to address the problem.

SUSTAINING A NATION

ABOVE: A dust storm blowing into Redcliffs, Victoria, from the Mallee wheat belt in the 1950s. By the 1930s, dust storms were commonplace in areas of soil erosion caused by the effects of overcropping, rabbits and drought. Winds would pick up the topsoil and blow it for miles. A dust storm could take hours to pass through towns, causing instant darkness and leaving a haze in the atmosphere. Planting clovers and farm consolidation that relieved the need to 'flog' small holdings eventually led to soil improvement.

ABOVE: Clearing palm scrub in the Ingham district, Queensland. Despite the rise of the conservation movement in the 1880s, most rural inhabitants, spurred by government policies and initiatives, were concerned with maximising the land's productivity. Clearing — which often involved the destruction of vegetation that had regenerated after earlier clearing in the 19th century — was fundamental to this goal.

LEFT: Sheep wearing bags for collecting faecal samples for worm egg counts in the 1940s. Such tests were carried out to monitor the efficacy of control measures including rotational grazing and drenching.

SUSTAINING A NATION

ABOVE: Landcare group members at a salinity control project site. Landcare, which aims to plant a billion trees, is one of the many programs aimed at rehabilitating degraded land and waterways.

SUSTAINING A NATION 133

PART THREE

SHOWCASE OF PARTICIPANTS

ROLL OF HONOUR

LEAD PARTICIPANTS

ABARE
The Dairy Farmers Group
National Farmers' Federation
Rural Press Limited

MAJOR PARTICIPANTS

Agriculture, Fisheries and Forestry — Australia
Australian Fisheries Management Authority
Australian Seafood Industry Council
CANEGROWERS
Consolidated Pastoral Company Pty Ltd
Consolidated Meat Group Pty Ltd
FreightCorp
Goodman Fielder
NSW Agriculture
NSW Forest Products Association Ltd
NSW Grains Board
Ricegrowers' Co-operative Limited
Sugar Research and Development Corporation
Sydney Fish Market Pty Limited
Sydney Markets Limited
Twynam Agricultural Group

Victorian Department of Natural Resources and Environment — Agribusiness Group
Victorian Department of State and Regional Development — Food Victoria
Westpac Banking Corporation

KEY PARTICIPANTS

Avcare Ltd
Aventis CropScience Pty Ltd
GrainCorp
Manildra Flour Mills Pty Ltd
MIA Council of Horticultural Associations Inc
Murray Goulburn Co-operative Co Limited
Queensland Rural Adjustment Authority (QRAA)
Ridley AgriProducts Pty Ltd

ABARE
Innovation in Economic Research
www.abareconomics.com

EXPANDING OUR AGRICULTURAL HORIZONS

ABN: 24 113 085 695

Core 6, Edmund Barton Building
Barton ACT 2601
GPO Box 1563
Canberra ACT 2601

Phone: 02 6272 2000
Fax: 02 6272 2001
Email: sales@abare.gov.au
Website: www.abareconomics.com

Dr Brian Fisher
Executive Director

BELOW: Women have played an important role in agriculture over the years.

ABARE, Australia's largest applied economic research agency, is an established name in circles of government, industry, business, research and the media. We analyse crucial economic issues affecting the commodities and resources sectors. And nowhere is this contribution more evident than in Australia's rural industries. But just as our rural sector has changed, so has ABARE evolved since its inception in 1945.

Before the Second World War, no government agency existed to advise on economic policy relating to the rural sector. The Royal Commission on Money and Banking in 1936 highlighted the need for research into economic issues in Australia. Some banks responded by establishing economic departments, such as that in the Rural Bank of New South Wales. The New South Wales Department of Agriculture set up a Division of Agricultural Economics in the following year.

When war broke out in 1939, the Commonwealth government had to divert many resources to the war effort. The Department of War Organisation of Industry was set up in 1941, which in turn established a Rural Industries Division in 1943 to undertake agricultural economics research.

Also in 1943 the government created the Ministry of Post-War Reconstruction. It included a Rural Reconstruction Section, which provided secretariate and research facilities for the Rural Reconstruction Commission. The commission found that economic considerations needed greater weight in agricultural policy. Thus the Rural Reconstruction Section expanded to become the Rural Research Division, and in April 1945 it amalgamated with the Rural Industries Division of the Department of War Organisation of Industry.

Meanwhile rural industry organisations were pressuring for ongoing economic research beyond the life of the Ministry of Post-War Reconstruction.

SUSTAINING A NATION

The ministry and the Department of Commerce and Agriculture made a joint cabinet submission in April 1945 to propose a Commonwealth Bureau of Agricultural Economics (BAE). The BAE (to become ABARE in 1987) was established as a division of the ministry in July 1945, to transfer to the Department of Commerce and Agriculture soon after the war ended.

THE FLEDGLING BUREAU — 1945–60 Two main tasks occupied the fledgling BAE. First, the War Service Land Settlement Scheme had high priority. The government was anxious to set up soldier settlers on viable 'home maintenance areas', and the BAE advised on the income earning potential of thousands of settlement proposals.

This work incorporated market outlook assessments for the industries covered in the proposals. These assessments were the start of the bureau's largest and most high profile activity — the ongoing provision of commodity forecasts.

The BAE's second major task was to help administer domestic pricing arrangements for agricultural commodities. The memory of the volatile commodity prices during the 1930s and wartime price controls led to pressures for postwar agricultural pricing policies. Such arrangements first covered the wheat and dairy industries, and the BAE helped collect the first farm surveys data for use in setting prices in these industries. This was the first step to ABARE's survey program today.

The other work of the BAE continued to reflect the changing economic environment, including the severe food shortages in Europe, postwar immigration to Australia, the outbreak of the Korean War, the wool boom of 1950–51, and the growing demand for imports to help the nation's reconstruction and development efforts.

Then in 1957 the Treaty of Rome set up the European Economic Community, and the BAE began analysing potential implications for Australia of Europe's Common Agricultural Policy. The BAE also published analyses of US farm policies, well before the chronic US farm surpluses emerged.

TOP: ABARE's predecessor the BAE advised on the income earning potential of thousands of soldier settlement proposals.

ABOVE: Storm over an irrigated rice crop, Leeton New South Wales. Irrigation and water policy reform have long been an important focus for ABARE research.

SUSTAINING A NATION

IDENTIFYING PROFITABLE AREAS FOR AGRICULTURAL EXPANSION — 1960s

The BAE was under pressure to conduct production oriented research to identify profitable areas for agricultural expansion. This work led to evaluations of land and water development proposals — for example, the BAE studies of the Ord river scheme.

By the end of the 1960s, Australian agricultural policy had swung from boosting rural development to dealing with the rural recession brought on by excess supplies in world commodity markets. Subsequent adjustment schemes in the dairy and horticultural industries were developed using the BAE's farm survey data and market outlook assessments.

FOCUSING ON THE NEEDS OF FARMERS — 1970s

Policy in the early 1970s was about facilitating farm adjustment and economic efficiency. The BAE published reports on farm debt and rural credit, contributing to the development of adjustment assistance measures.

The BAE held its first national agricultural outlook conference in 1971. The conference was a response to the information needs of farmers, industry leaders and government, who were trying to cope with a severe downturn in the rural economy. An annual event, the conference has grown in scope and public accessibility. And today it is the pinnacle of ABARE's public presence.

RURAL SECTOR REFORM — 1980s and 1990s

The 1980s were a time of reform for Australia's rural sector. And there is now less direct government intervention in the affairs of our rural industries.

Assistance levels are much higher in foreign industries with which Australia competes. Thus a major thrust of research in the 1980s and 1990s was to analyse such assistance and its costs. The results of this research have been available to Australian negotiators seeking to

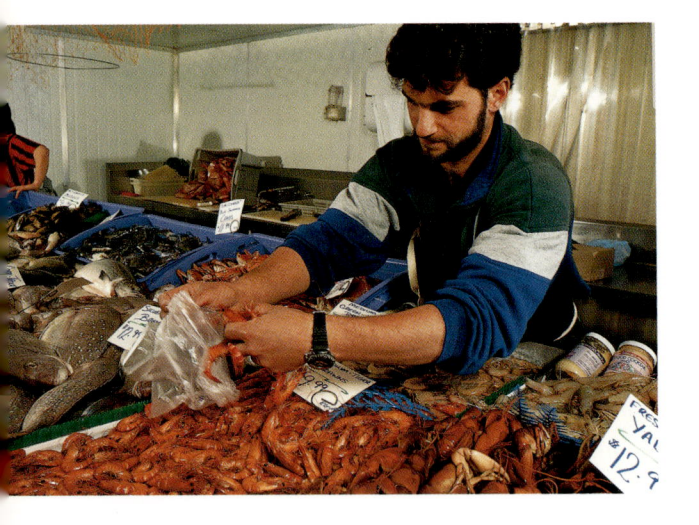

ABOVE: In the 1990s ABARE undertook extensive economic analysis of forest resource use options which contributed to the government's Regional Forest Agreement process.

TOP: An important role for ABARE has been the development of an economic database on Australian fisheries.

influence policy in the European Union, the United States and Japan especially. They underpinned Australia's contribution to the Uruguay Round of multilateral trade negotiations, for example. ABARE has maintained this research on Australia's emerging sphere of interest in the Asia Pacific region.

After expanding into new areas of research in the 1970s (the forest and wood products industries) and 1980s (fisheries), the organisation accelerated its involvement in research into natural resource management issues. A 1987 departmental reorganisation, merging the BAE and the Bureau of Resource Economics (BRE), opened the research horizons of the bureau.

The newly named ABARE incorporated new commodities (including minerals and energy) in its market outlook analysis and research. It also became involved in resource assessments, analyses of resource use conflicts, and environmental research issues.

ABARE TODAY Not unlike the BAE set up 55 years ago, ABARE is still largely in the business of adding value to the information it collects, analyses and disseminates. Its research focuses on:

- commodity outlook forecasts
- farm performance
- world agricultural trade analysis
- water market reform
- international energy trade analysis
- forestry management
- climate change policy analysis
- global modeling
- natural resource management
- energy market reform

For its clients in both the private and public sectors, ABARE continues to provide information on economic issues that affect the management of our basic assets — land, water, forests, fisheries, minerals, energy resources and the environment.

TOP: Deregulation and adjustment of the dairy industry has been an important issue for ABARE in the 1980s and again today.

ABOVE: The wool industry has figured prominently in ABARE's research from the days when 'the country rode on the sheep's back' till the current day.

SUSTAINING A NATION

SUSTAINABLY MANAGING OUR FISHERIES AND PROTECTING THE MARINE ENVIRONMENT

ABN: 24 113 085 695

Macquarie Street
Barton ACT 2600
GPO Box 858
Canberra ACT 2601

Phone: 02 6272 3933
Fax: 02 6272 5161
Email: pr.contact@affa.gov.au
Website: www.affa.gov.au

Mr Michael Taylor
Secretary

BELOW: Our fisheries operate in one of the world's cleanest environments.

AUSTRALIA HAS THE world's third largest fishing zone (9 million square kilometres) and Australian fisheries operate in one of the world's cleanest environments. In 1998–99, the gross value of fisheries production was estimated at $2.04 billion. This included $602 million from aquaculture — one of Australia's fastest growing rural industries.

The Department of Agriculture, Fisheries and Forestry Australia (AFFA) is the Commonwealth agency with responsibility for Australia's agricultural, fishing, forestry and processed food industries. As well as taking a lead role in the research and development of sustainable fisheries policies, AFFA is involved in a range of other initiatives including:

- the negotiation of arrangements for access to, and for the sustainable management of, shared and migratory fish stocks in international waters;
- the negotiation of arrangements to deal with associated illegal, unreported and unregulated fishing activity in international waters;
- the development of national arrangements to prevent and respond to disease outbreaks in aquatic animals (AQUAPLAN) and the invasion of exotic marine pests;
- fish habitat protection, restoration and awareness raising through the Fisheries Action Program;
- negotiations with foreign governments to increase market access and trade for Australian products;
- policy development for the management and development of Australian domestic Commonwealth fisheries;
- the development of an Action Agenda for Aquaculture, to facilitate the sustainable development of the industry.

SUSTAINING A NATION

MANAGING AUSTRALIA'S FISHERY RESOURCES

THE AUSTRALIAN FISHERIES MANAGEMENT AUTHORITY (AFMA) is the statutory body responsible for sustainably managing the Commonwealth's fishery resources with a strong emphasis on a cooperative partnership between all parties from fisheries managers and commercial operators to scientists, environmentalists and recreational fishers. AFMA believes that to achieve sustainable fisheries there must be a level of confidence and trust.

Since its inception in 1992, AFMA has emphasised the importance of all stakeholders taking both ownership of decisions and greater responsibility for the wellbeing of individual fisheries. This successful model is now being adopted by other agencies within Australia and internationally.

AFMA manages fisheries within the 200 nautical mile Australian fishing zone, developing appropriate policies to promote a sustainable, economically efficient, commercial industry.

There is a commitment to providing fishers with secure and transferable rights to enable the efficient harvesting of commercial species. Individual transferable quotas are the preferred method of managing Commonwealth fisheries.

The Authority operates in a transparent manner, regularly consulting with industry through management advisory committees established in each major fishery. An eight-member expertise-based Board is responsible for overseeing AFMA's operations, setting the Authority's policy framework and making high level decisions based on advice and recommendations from its sub-committees and MACs.

The Authority also has a key role in the detection and investigation of illegal activities by domestic and foreign fishing boats. AFMA undertakes this function in conjunction with other Commonwealth and State agencies, providing overall coordination, policy direction, technical advice and funding.

ABN: 81 098 497 517

3rd Floor, John Curtin House
22 Brisbane Avenue
Barton ACT 2600
Box 7051, Canberra Mail Centre
Canberra ACT 2610

Phone: 02 6272 5029
Fax: 02 6272 5036
Email: info@afma.gov.au
Website: www.afma.gov.au

Mr Stuart Richey
Acting Chairman

Mr Frank Meere
Managing Director

SUSTAINING A NATION

HELPING CHART A SECURE FUTURE FOR THE AUSTRALIAN SEAFOOD INDUSTRY

ABN: 35 008 664 999

Unit 1, 6 Phipps Close
Deakin ACT 2600
PO Box 533, Curtin ACT 2605

Phone: 02 6281 0383
Fax: 02 6281 0438
Email: asic@asic.org.au
Website: www.asic.org.au

Mr Nigel Scullion
Chairman

Mr Russ Neal
Chief Executive Officer

BELOW: Seafood — helping to build the nation's economy.

AUSTRALIAN SEAFOOD is renowned worldwide for its high quality product drawn from clean waters. As Australia celebrates the new millenium and the Centenary of Federation, the nation's gross value of seafood production has broken through the $2 billion barrier. The industry has excellent prospects for growth on both the export and domestic markets.

At a national level, the industry is represented by the Australian Seafood Industry Council (ASIC). This is the peak body for the Australian fishing, aquaculture and post-harvest seafood industries. In its advocacy role, ASIC is committed to pursuing issues and developments that enhance seafood sustainability and profitability. ASIC plays a vigorous, effective role in the development of policies and reforms to benefit the industry and help chart a secure future.

The industry has organisations representing each state or specific fisheries. The ASIC Board comprises seafood industry councils from Queensland, New South Wales, Victoria, Tasmania, South Australia, Western Australia and the Northern Territory, specific fisheries including the East Coast Tuna, Southern Bluefin Tuna, South East Trawl and Northern Prawn fisheries, and the Master Fish Merchants Association.

ASIC welcomes enquiries on any aspect of Australian seafood production, processing and distribution.

SUSTAINING A NATION

A PRO-ACTIVE APPROACH TO PROVIDING A CLEANER ENVIRONMENT

AVCARE LTD, the association representing agricultural and veterinary chemical manufacturers and distributors, takes a pro-active approach to environmental management.

Avcare executive director Claude Gauchat said the $1.8 billion industry was responding to heightened community expectations. 'Our members are an integral part of the rural scene and long-term sustainable agricultural production is very much in their interests.' A co-regulatory code of practice program operates throughout the distribution supply chain managed by a subsidiary company Agsafe. Training is provided in the safe storage, handling transport and sale of agricultural and veterinary chemicals with accreditation for both staff and premises. 'Agsafe has authorisation granted by the ACCC on behalf of industry, to levy trading sanctions against non compliance,' Claude Gauchat said.

To address the problem of empty chemical containers accumulating around the countryside, Avcare was instrumental in the concept of *drumMUSTER*, launching the program in 1999 with partners the National Farmers Federation, the Veterinary Manufacturers and Distributors Association and the Australian Local Government Association. This multi-stakeholder program provides for regular collection of metal and plastic non-returnable containers for recycling.

Avcare has also been involved in a state and federal government waste management initiative which is proving successful in the west and is about to be adopted nationally. ChemCollect's main target is organochlorine pesticide waste.

Following completion of the government's ChemCollect program, industry has agreed to implement the Chemclear program that will collect and dispose of unused unwanted registered agricultural and veterinary chemicals stored on farms. 'These practical initiatives demonstrate the commitment of the chemicals industry to providing a cleaner and safer environment,' Claude Gauchat said.

ABN: 29 008 579 048

Level 2, AMP Building, 1 Hobart Place
Canberra ACT 2601
Locked Bag 916
Canberra ACT 2601

Phone: 02 6230 6399
Fax: 02 6230 6355
Email: avcare@avcare.org.au
Website: www.avcare.org.au

Mr Mark Allison
President

Mr Claude Gauchat
Executive Director

BELOW: *drumMUSTER* — rinse them out, round them up and run them in.

SUSTAINING A NATION

Aventis

COMMITTED TO PRODUCT INNOVATION WHICH ADDS VALUE IN THE PADDOCK

ABN: 87 000 226 022

391-393 Tooronga Road
Hawthorn East VIC 3123

Phone: 03 9248 6888
Fax: 03 9248 6800
Email: carolyn.key@aventis.com
Website: www.aventis.com.au

Mr Olivier Duroni
Managing Director

Mr David Cashmore
Finance/Admin/IT Manager

AVENTIS CROPSCIENCE boasts a proud global history of more than 150 years in agricultural markets.

The company, formed through the international merger of Hoechst AG and Rhone-Poulenc SA, enjoys market leadership in crop production in more than 160 countries around the world.

Aventis Australia's Managing Director, Olivier Duroni, said Aventis is committed to working in partnership with stakeholders to deliver better products and formulations from seed selection to harvest.

'We have expanded resources and expertise to make a crucial contribution to sustainable agricultural production in this country. Aventis staff and our distributors are able to provide farmers with technologically advanced information on our products and services and expert advice on their use.

'We are committed to product innovation which adds value in the paddock,' Olivier Duroni said.

BELOW: The outside of canvas water bags was smeared with tallow.

ABOVE AND RIGHT: Aventis — bringing new ideas to agriculture.

SUSTAINING A NATION

SUSTAINING A NATION

THE FAMOUS WINEGLASS BRAND — A WORLDWIDE SYMBOL OF QUALITY BEEF

ABN: 22 010 080 654

Newcastle Waters Station
Newcastle Waters NT 0862

Phone: 08 8964 4527
Fax: 08 8964 4533
Email: admin@pastoral.com
Website: www.pastoral.com

Mr Ken Warriner
*Managing Director and
Chairman of the Board*

BELOW: CPC has one of the largest breeding herds in Australia.

ONE OF AUSTRALIA'S major beef industry players, Consolidated Pastoral Company (CPC), has emerged in less than 20 years.

The group, which has one of the largest breeding herds in the country, is a division of Mr Kerry Packer's Consolidated Press Holdings Ltd.

CPC was formed in 1983 with the purchase of Newcastle Waters Station and some smaller holdings in the Northern Territory. Since that time, strategically located properties have been added to the group across Western Australia and the Territory, through Queensland and New South Wales. Concentration on the vast grazing areas across northern Australia has enabled successful drought management and marketing flexibility with easy access to export markets for quality grassfed and lotfed beef and live cattle.

The company has extended its holdings to 17 pastoral properties totalling more than 44,000 square kilometres. With the 1993 purchase of the international agricultural consulting and management company GRM International, Austrex, a company specialising in the export of live cattle around the world and Consolidated Meat Group (see following page), CPC became responsible for the management of more than 300,000 cattle across northern Australia.

CPC has been a force in improving the genetic makeup of the northern beef herd with three large Brahman studs operating at Argyle Downs in Western Australia, Newcastle Waters in the Territory and Allawah in Queensland in addition to a Charolais stud at Mimong Station, Queensland.

The famous wineglass brand that originated on Newcastle Waters more than a century ago now appears on CPC cattle throughout Australia and has become internationally recognised as a symbol of quality beef.

SUSTAINING A NATION

A COMMITMENT TO SUPPLYING PREMIUM BEEF

CONSOLIDATED MEAT GROUP
Australia's Finest Meat to the World

THE LARGEST wholly Australian-owned beef enterprise, Consolidated Meat Group (CMG), is fittingly based in the nation's beef capital, Rockhampton in central Queensland.

CMG, a key rural business unit owned by Mr Kerry Packer's Consolidated Press Holdings Ltd, which produces over 100,000 tonnes of beef a year for markets throughout Australia and around the world, began operating in 1994 as a joint venture between Consolidated Pastoral Company and Tancreds. The Tancred family, respected meat industry operators for four generations, has subsequently sold their interest to Consolidated Pastoral Company.

Rockhampton is the location of CMG's flagship Lakes Creek abattoir, which has undergone multi-million dollar upgrading to become one of the largest beef processing plants in Australia. Other modern abattoirs located at Katherine in the northern Territory and Innisfail in north Queensland give a combined processing capacity of more than 3000 cattle a day. All three abattoirs exceed the stringent standards required by European and United States regulators and have Halal accreditation.

The commitment to supplying premium beef extends through strategic alliances back to the property level with CMG embracing the national on-farm quality assurance program CattleCare. All cattle processed are individually identified. Beef is processed and marketed to strict specifications for the diverse international customer base across a broad range of fresh meat cuts with increasing emphasis on value added products. A rigorous program of microbiological testing in association with AQIS is in place at all three abattoirs to guarantee product safety and integrity.

CMG enjoys a significant share of the Australian beef market, supplying customers through distribution centres in Sydney, Melbourne, Brisbane and Adelaide.

ABN: 65 065 093 709

Private Mail Bag
Rockhampton QLD 4700

Phone: 07 4930 5888
Fax: 07 4930 5839
Email: cmg.info@ausmeat.com
Website: www.meat.aust.com

Mr Ray O'Dell
Executive Chairman

Mr Brian McFarlane
Marketing Manager

BELOW: CMG supplies premium beef worldwide.

SUSTAINING A NATION

DAIRY FARMERS — A NATIONAL ICON

ARBN: 010 308 068

168 Walker Street
North Sydney NSW 2060
PO Box 1810
North Sydney NSW 2059

Phone: 02 9903 5222
Fax: 02 9957 3449

Mr Ian Langdon
Chairman

Mr Alan Tooth
Managing Director

BELOW: Hand milking was common on dairy farms until the 1950s.

THE RAPID CHAIN of events that has transformed the Dairy Farmers Group into the largest dairy enterprise in Australia spans barely a decade.

Yet the business which now boasts annual turnover exceeding $1.3 billion can trace its origins to a group of 70 pioneering New South Wales producers who banded together to form a milk supply co-operative in 1900 as Australian Co-operative Foods Ltd.

In the year 2000, Dairy Farmers can proudly say it has been contributing to Australia's development for 100 years and predates Federation.

The fledgling Dairy Farmers Co-operative had to have ice and transport so all its start-up capital was poured into purchasing a five-tonne refrigerator, milk tanks, horses and carts. It took a decade of hard work before the co-operative was in profit in 1912 and able to pay its first dividend.

There were many milestones along the way including the first bottled milk in 1925, the launch of a free school milk program in 1941 and the production of Australia's first long life milk in 1977.

But the modern day revolution began only in 1989.

Australian Co-operative Foods was born that year from a merger of the three long-established NSW co-operatives; Dairy Farmers, Hunter Valley and Shoalhaven.

Through a series of further mergers and acquisitions, the Dairy Farmers Group has emerged as a national food icon, supplying both fresh and processed dairy produce throughout Australia and with a significant export market.

The quantum leap that took Dairy Farmers beyond the New South Wales border for the first time came in 1996 with the merger of Australian Co-operative Foods and the large Queensland-based co-operative Queensco-Unity Dairyfoods. Far northern processor Malanda Dairyfoods joined the following year.

SUSTAINING A NATION

In an historical move to 'buy back the farm' Dairy Farmers acquired the natural cheese business, including the Coon brand, from multi-national Kraft Corporation in December 1997.

The South Australian-based Dairy Vale Co-operative added to the strength of the Dairy Farmers flagship in 1998.

A key international link was established in January 1999 when Group Danone, France's leading food and beverage company, entered into an alliance to produce Danone yogurts and dairy desserts.

The alliance includes business development in Asia where Danone is strongly established and Dairy Farmers has growing markets.

Group exports have grown at a phenomenal rate. From a sales base of less than $1 million in 1993, overseas trading is now worth more than $114 million. Today Dairy Farmers exports a range of quality products by sea and air freight to 45 countries including Singapore, Hong Kong, Japan and the Middle East. Exported products include cheddar, mozzarella, sweetened condensed milk, icecream and milk products.

The popular Ski brand of yogurts has established market leadership in Singapore.

Fresh, powdered, condensed and UHT milk together with icecream, cream cheese, cheddar and casein are the mainstays of the Dairy Farmers extensive range of exports.

ABOVE LEFT AND RIGHT: Dairy Farmers has maintained an ongoing presence at major agricultural shows over the years.

SUSTAINING A NATION

The original band of farmer shareholders has now swollen to more than 5500, spread from the Atherton Tableland in far north Queensland to the Adelaide Hills in South Australia.

The Group boasts a string of market-leading retail brands which are household names across the country. These include Bornhoffen, Coon, Cruise, Dairy Farmers, Dairy Vale, Eve, Farmers Best, Fred Walker, Jacaranda, Mil Lel, Oak, Shape, Malanda and Unity.

Brands manufactured under license include Cadbury, Chill, Cracker Barrel, Danone, Ski and Weight Watchers.

Dairy Farmers is now the national leader in table milk with more than 35 per cent of the market, in yogurt with 35 per cent and in natural cheese with 25 per cent. In the foodservice market, it is also the market-leading supplier of mozzarella cheeses to major pizza chains.

The Group employs about 3000 workers at 26 locations in city and country. Farmer shareholders supply Australia with almost 1.4 billion litres of milk a year. Of this total, about 800 million is litres turned into value added dairy products.

After a century of attention to quality and innovation the Dairy Farmer Group now proudly claims the mantle of Australia's Dairy Best.

An extensive promotion campaign built around this theme was launched in New South Wales and rolled out in Queensland, Victoria and South Australia to achieve a single, consistent branding across all markets.

Dairy Farmers also has a strong commitment to community and sporting sponsorship. Programs range from school athletic clinics to scholarships for emerging stars to major backing for women's netball.

A reputation for product excellence and response to changing consumer demands has the Dairy Farmers Group well placed to tackle the challenges of a deregulated dairy market in Australia.

TOP: Four-wheeler fitted with pneumatic rubber tyres for suburban bottle delivery.

ABOVE: As transport improved, methods of milk delivery became more sophisticated.

Under a regime of partial deregulation in New South Wales and Queensland, Dairy Farmers has achieved outstanding growth in market share of fresh milk sales. Potential for further growth in the consumption of dairy products across the board is being realised with a market focus on a healthy lifestyle and a balanced diet.

The shape of milk marketing was changed significantly in 1998 with the introduction of the first user-friendly three-litre screw top bottle which consumers have voted a winner.

The company which also pioneered the manufacture of long-life milk has now taken product marketing a step further with a dedicated $22 million facility producing Australia's first UHT milk in screw top plastic bottles.

Dairy Farmers has been a leader in the development of low fat modified milks and was the first in the world to produce a no cholesterol milk.

Mil Lel Superior cheeses have been prolific gold medal winners at the prestigious Royal Melbourne and Sydney dairy shows and have won gold, silver and bronze at the World Cheese Championships in the USA.

Dairy Farmers reputation for product innovation has been enhanced by winning Foodweek Magazine's Most Successful New Product award for three successive years with Farmers Best cholesterol free milk, Ski Double Up and Cadbury Mousse Dessert.

A key to Dairy Farmers product handling integrity and close customer relationship is its unique vendor system. To better meet the challenges of a deregulated milk market, former vendor networks have been restructured into a dedicated team of more than 1200 'milkos' who deliver the extensive product range direct to retail outlets, food stores and households. Each vendor is a franchised owner whose local territory experience is a valuable asset in a competitive marketplace.

The Dairy Farmers Group has invested more than $60 million in capital expansion of its facilities around the country in the past year and is poised to reap the benefits of future market growth.

TOP: In the mid 1950s, bottling and crating milk for distribution.

ABOVE: Product testing at Toowoomba, Queensland — an important part of quality assurance.

FreightCorp

DELIVERING A COMPETITIVE, COST-EFFECTIVE TRANSPORT AND LOGISTICS SERVICE

ABN: 97 953 673 525

Level 13, Jessie Street Centre
2–12 Macquarie Street
Parramatta NSW 2150

Phone: 02 9893 2500
Fax: 02 9893 2501
Email: corporateaffairs@freightcorp.com.au
Website: www.freightcorp.com.au

Mr Barry Capp
Chairman

Mr Lucio Di Bartolomeo
Managing Director

BELOW: Horses remained an important form of transport until the 1930s.

AUSTRALIAN SYDNEY-BASED rail operator FreightCorp, has thrived in a competitive deregulated environment, hauling a record 86 million tonnes of freight in the past year.

FreightCorp — the trading name of the Freight Rail Corporation — has made giant strides since corporatisation in July 1996, building on and expanding its core business of hauling grain and coal in New South Wales. The implementation of an aggressive program to establish itself as a major player in the national transport market has seen an extension of services into Victoria, Queensland and South Australia, and expansion into intermodal freight, customised and packaged transport solutions, terminal management and container storage.

'We have done the hard yards and are now positioned to compete successfully and establish ourselves for the long term as one of the premier providers of rail-based transport services to the national bulk and general freight markets,' said Managing Director Lucio Di Bartolomeo.

On-going cost savings were being achieved with an average 25 per cent reduction in freight rates passing on a more than $150 million benefit to customers since corporatisation.

Lucio Di Bartolomeo said the cost reductions were particularly important for the grain and coal sectors, export industries under pressure from falling prices on world commodity markets. Grain and coal account for more than 90 per cent of FreightCorp business, with coal haulage volumes from the Hunter Valley to the port of Newcastle now at over 66 million tonnes a year.

In 1999–2000, FreightCorp successfully coordinated delivery of a record-breaking NSW grain harvest, meeting the challenge of direct competition for market share. The corporation plays a key role in supply chain management with primary producers and customers — for example, integrating the logistics

SUSTAINING A NATION

of product movement from grain silo to flour mill and through starch/gluten manufacture to export through Port Botany for the Manildra Group.

The transport of containerised general freight has grown significantly under the FreightCorp PortLink strategy with volumes nearly doubling to 150,000 TEUs since 1996. Strategic alliances with producers and exporters of bulk and containerised agricultural product have contributed to growth.

Lucio Di Bartolomeo said that while operating practices and systems had been re-engineered to achieve cost effectiveness and a higher quality service, the pressure to find continuing efficiencies would remain.

In December 1998, FreightCorp moved outside its New South Wales base and secured a major 10-year contract to haul coal for Flinders Power in South Australia. Intermodal services have since been extended into Melbourne and Brisbane. In March 2000, FreightCorp was selected by a group of Queensland companies as preferred tenderer for a 10-year contract to haul 6 million tonnes per year of domestic and export coal in Central Queensland — FreightCorp's first operation on a narrow gauge network.

Lucio Di Bartolomeo said the experience gained by FreightCorp over the last four years left the organisation well placed to play a leading role in reshaping the rail industry in Australia.

TOP: FreightCorp plays a major role in coordinating the movement of the NSW grain harvest.

ABOVE: Fast turnaround at Port Botany has been achieved through cooperative work with stevedores and customers.

Goodman Fielder

A FOOD COMPANY THAT'S FAST BECOMING A GLOBAL FORCE

ACN: 000 003 958

75 Talavera Road, Macquarie Park
North Ryde NSW 2113
Locked Bag 2222
North Ryde NSW 2113

Phone: 02 8874 6000
Fax: 02 8874 6099
Email: corporate.affairs@goodmanfielder.com.au
Website: www.goodmanfielder.com.au

Mr David Hearn
CEO and Managing Director

Mr Jon Peterson
Chairman

BELOW: Wheat harvesting at Horsham in Victoria.

GOODMAN FIELDER, Australasia's largest food company, has its sights set firmly on achieving recognition as a world-class player. From its newly consolidated Sydney headquarters, operations reach into more than 40 countries with a workforce which has grown to around 18,000 employees.

Goodman Fielder operates across the retail, food service and industrial sectors and is best known for its grain-based products, including breads, cereals, margarines and cooking oils and food ingredients.

Brands that enjoy a loyal consumer following include Uncle Toby's, Buttercup, Meadow Lea and White Wings. Goodman Fielder became the only national milling and baking company in Australia with the 1999 purchase of Bunge Defiance bringing into the company the Defiance, Sunicrust and Country Bake labels.

Its position in the New Zealand market has been strengthened with the acquisition of Ernest Adams, extending market share for baked foods including pastries and cakes.

Another strategic acquisition in the Chinese premium edible oil market has come with the 90 per cent acquisition of Shanghai Van Den Bergh.

Goodman Fielder Chief Executive David Hearn said the major corporate changes were essential building blocks for successful operation in the challenging food industry environment.

Market research and innovation has allowed Goodman Fielder to keep pace with food trends. After 18 months of intensive research and product development, a new range of restaurant quality convenience meals has been launched by GF Fresh.

To meet consumer lifestyle changes, Meadow Lea now comes in salt reduced and low fat versions and Uncle Toby's has released fat reduced foods,

SUSTAINING A NATION

such as Lite Start, a folate and Vitamin E enriched low fat breakfast cereal. Folate is also present in Uncle Toby's VitaGold bread.

The health benefits of so-called 'functional foods' — foods that play a potential role in disease prevention, mitigation and cure — offer a number of opportunities. Goodman Fielder is already responding to these trends with products such as Gold'n Canola, which contains the Omega Benefit, and Uncle Toby's Lite Start. Goodman Fielder has proven innovation skills in developing healthy products such as Hi-Maize, the key ingredient of Wonder White.

The nutritional value and safety of food are becoming critical issues for many people. Goodman Fielder has acted on this concern, ensuring its 'Good Food Guaranteed' trust mark appears on all of the company's products.

David Hearn said Goodman Fielder was committed to a wide-ranging R&D program in Australia and overseas.

Goodman Fielder has committed to a range of community and sporting sponsorships including one of the largest programs ever undertaken as a key food supplier to the Sydney Olympics.

An agreement has been signed with Foodbank, a non-profit organisation that distributes surplus products from manufacturers to welfare agencies.

The company participates in the National Packaging Covenant, a voluntary national program that aims to reduce the amount of packaging waste going into landfill.

Goodman Fielder has backed a public education campaign by the environmental group Landcare Australia, encouraging sustainable use of natural resources.

The company is also forging links with rural industry through sponsorship of the Australian Rural Leadership Program.

TOP: Allied Flour and Baking Services manufacture a range of quality flours and meals for the baking and food manufacturing industries Australia-wide.

ABOVE: A canola field in Wagga Wagga, New South Wales.

SUSTAINING A NATION

GrainCorp
The Grain Growers Company

INNOVATION AND FLEXIBILITY TO MEET GRAIN CUSTOMER REQUIREMENTS

ABN: 52 003 875 401

Level 10, 51 Druitt Street
Sydney NSW 2000
PO Box A268
Sydney South NSW 1235

Phone: 02 9325 9100
Fax: 02 9325 9180
Website: www.graincorp.com.au

Mr Ron Greentree
Chairman

Mr Tom Keene
Managing Director

BELOW: A welcome break in the early days of harvesting.

CHANGES IN THE Australian grains sector are viewed as an opportunity by leading agribusiness service company GrainCorp. The traditional grain silo operator in NSW is a thriving enterprise which is meeting head-on the challenges posed by globalisation, deregulation and biotechnology.

'Our business is increasingly focused on adding value for our producers and customers. We're building a grain solutions culture built on innovation and flexibility in meeting grain customer requirements,' said GrainCorp Managing Director, Tom Keene.

GrainCorp operates principally in New South Wales and Victoria. It also operates a $7.3 million storage and handling complex at Goondiwindi, Queensland. GrainCorp has joint venture interests in the biotechnology company Nugrain Ltd and has shared investment in new storage facilities with multi-national grain trader Cargill Ltd.

ABOVE: Each year GrainCorp handles an average 6 million tonnes of Australian grain.

SUSTAINING A NATION

A MAJOR FORCE IN AUSTRALIA'S AGRICULTURAL AND FOOD INDUSTRIES

A COMMITMENT TO EXCELLENCE has been the business foundation of the Manildra Group, a family-owned enterprise that has diversified from a flour milling base into industrial wheat and sugar processing in Australia and the United States.

From the purchase of a country flour mill at Manildra in New South Wales in 1952, the company has grown and expanded to become a specialist supplier to domestic and international markets.

The sale of flour and pre-mixes has been complemented by a growing range of wheat-based products including starches, gluten, fructose and glucose syrups.

The group now boasts specialised processing facilities at Nowra, south of Sydney with a second modern flour mill at Gunnedah in northern New South Wales.

Manildra has pioneered international markets for protein products and established a US subsidiary marketing company before setting up its own starch and gluten manufacturing facilities in Minneapolis and Minnesota. The company also operates an ethanol plant in Iowa.

Manildra has built a reputation for the supply of specialised starches to the paper and cardboard manufacturing industry. In 1997 a purpose-built starch research facility was opened in Melbourne with expert staff to trial and develop new products and applications.

For more than a decade, Manildra has been a major player in the sugar market, operating a joint venture refining and marketing company with the New South Wales Sugar Milling Co-operative.

ABN: 80 000 217 523

PO Box 72, Auburn NSW 1835

Phone: 02 9649 1444
Fax: 02 9646 4619
Email: manildra@manildra.com.au
Website: www.manildra.com.au

Mr Dick Honan
Chairman

Mr Peter Simpson
General Manager

Mr John Honan
Marketing Manager

BELOW: With the introduction of bulk handling, wheat was carted to depots.

SUSTAINING A NATION

MIA Council of Horticultural Associations

PROMOTING LONG-TERM SUSTAINABLE IRRIGATED HORTICULTURE IN THE MIA

20 Olympic Street
Griffith NSW 2680
PO Box 1059
Griffith NSW 2680

Phone: 02 6964 2420
Fax: 02 6964 2409
Email: miacha@webfront.net.au

Mr Harley Delves
Chairman

Ms Belinda Wilkes
Chief Executive Officer

BELOW: Wine grape vineyard in the Murrumbidgee Irrigation Area.

AS ONE OF AUSTRALIA'S PREMIER FARMING REGIONS, the Murrumbidgee Irrigation Area is a household name. The MIA Council of Horticultural Associations is the peak body promoting and protecting the area's industries.

CITRUS The MIA citrus industry produces on average 190,000 tonnes of citrus annually, for export, domestic and processing markets. Varieties produced include valencias, navels, grapefruit, mandarins, lemons and limes. The MIA comprises of 620 registered growers with approximately 9000 hectares of plantings.

WINE GRAPES The region produces 65 per cent of all wine grapes grown in NSW, 15 per cent of Australia's total production. Farm gate value is in excess of $100 million. Production is likely to exceed 200,000 tonnes in the future from 12,000 hectares of plantings and 550 growers.

PRUNES The region produces approximately 2000 dried tonnes of prunes — about 60 per cent of the New South Wales total, with a farm gate value of $4 million for the domestic market and export. Around 500 tonnes of high value, fresh sugarplums are grown, mainly for export to Hong Kong, Singapore and Taiwan.

ABOVE: Citrus fruit (left) and prune trees (right).

SUSTAINING A NATION

MURRAY GOULBURN CO-OPERATIVE CO. LIMITED

Devondale
WHOLLY AUSTRALIAN

100 PER CENT AUSTRALIAN OWNED, 100 per cent Australian made. From humble beginnings in Cobram in 1950, Murray Goulburn has developed into Australia's largest exporter of processed food, processing around 30 per cent of the nation's milk.

The Co-operative is owned by 3500 dairy farmers principally in Victoria and spreading into southern New South Wales and south eastern South Australia. Milk is processed in six modern manufacturing plants situated across rural Victoria. With some 70 per cent of its 440,000 tonne production exported last year, Murray Goulburn is a major force on the global scene, accounting for about 6 per cent of world trade in dairy products.

The company has performed strongly despite world prices being at a low ebb with operating revenue in the past year growing to a record $1.3 billion.

ABN: 23 004 277 089

140 Dawson Street
Brunswick VIC 3056
PO Box 4307
Melbourne VIC 3001

Phone: 03 9387 6211
Fax: 03 9387 5741
Email: mgc@mgc.com.au
Website: www.devondale.com

Mr Ian MacAulay
Chairman

Mr Stephen O'Rourke
Managing Director

ABOVE: State-of-the-art milk powder processing at Murray Goulburn's Koroit plant.

BELOW: Shipping was important before the extension of road and rail transport.

SUSTAINING A NATION

AUSTRALIAN FARMERS — SUSTAINING THE NATION

ABN: 77 990 310 600

Level 3, NFF House
14-16 Brisbane Avenue
Barton ACT 2600
PO Box E10, Kingston ACT 2604

Phone: 02 6273 3855
Fax: 02 6273 2331
Email: nff@nff.org.au
Website: www.nff.org.au

Mr Ian Donges
President

Dr Wendy Craik
Executive Director

BELOW: Farming is a way of life as much as a livelihood.

THROUGHOUT AUSTRALIA'S DEVELOPMENT, farmers have played a prominent and vital role — laying the foundations for our strong economy and helping to build communities across the country.

Farmers have been involved in all the structures that make Australia what it is today — from the organisation and operation of our society; the industrial framework in which we work; the political landscape; and, most importantly, in our national wealth and economic stability.

Since July 1979 the National Farmers' Federation (NFF), based in Canberra, has represented the national and global interests of Australia's 120,000 rural producers.

As the peak farm organisation, the NFF represents producers of all major agricultural products through its membership base of independent state farm organisations, national commodity councils and affiliated groups.

The NFF was born out of the desire of producer groups to speak with one voice — farmers realised that the most effective means to a strong national influence was a single national entity.

Reflecting upon the many achievements by the NFF over 21 years, a common theme emerges — reform. NFF has built a professional and credible reputation as a national leader of reform and debate. Initially, this was focused in the economic and industrial area, later broadened to include the environment and, more recently, covering community service issues like health, education and telecommunications.

The reform process started with the threat to Australia's developing live sheep export trade in late 1979 and continued through a series of watershed events which helped shape the nation's landscape.

SUSTAINING A NATION

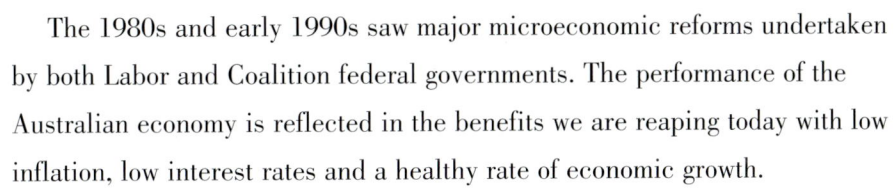

The 1980s and early 1990s saw major microeconomic reforms undertaken by both Labor and Coalition federal governments. The performance of the Australian economy is reflected in the benefits we are reaping today with low inflation, low interest rates and a healthy rate of economic growth.

A run through the financial health checklist is encouraging. Exports have remained strong and continue to diversify, particularly in agriculture and processed food and fibre; and a lower dollar is helping make us internationally competitive. But there are still many issues for the NFF to pursue to represent the national and global interests of Australia's 120,000 rural producers.

Farmers have historically been price takers, not price setters, so the priority is always going to be to get the settings right. The continuing decline in commodity prices means that the NFF will have most success by focusing on reducing farm costs in areas like industrial relations, taxation and by increasing market access.

Taxation reform has been a major issue for the Federation, which adopted a policy supporting a broad-based consumption tax more than a decade ago.

ABOVE LEFT: NFF has pushed hard for rural service upgrades and telecommunication facilities which will enable rural Australia to participate in the global economy.

ABOVE: The continuing decline in commodity prices means that the NFF will have most success by focusing on reducing farm costs in areas like industrial relations and taxation, and by increasing market access.

SUSTAINING A NATION

ABOVE: Labor market and industrial reforms remain at the top of the NFF agenda.

The introduction of the tax reform package, including the GST even if it is an imperfect model, is welcomed for the nearly $1 billion it promises to remove from farm costs.

Trade reform is another seminal issue for the NFF. Farmers export 80 per cent of their production and there are potentially huge benefits that can flow to Australian producers. As one of the leaders in the Cairns Group of free trading nations, Australia simply can't afford to sit back and wait for the World Trade Organisation to tackle the protectionist agricultural policies of Europe and North America. We have no choice — our economy cannot match their subsidised regimes.

Labor market and industrial reforms remain at the top of the NFF agenda. Despite major successes over the past 21 years there are significant efficiencies still to be achieved. NFF has featured prominently in issues like contract labor at abattoirs (Mudginberri in the Northern Territory); waterfront reform (live sheep, bulk grain and container loading); the use of wide combs in shearing; and the removal of the tally system in meat processing.

Many of the issues pursued by the NFF are likely to establish a national precedent. Support for such issues is sometimes provided through the Australian Farmers' Fighting Fund, launched at the 1985 farm rally in Canberra. The AFFF reached its capital target of $10 million with 40 per cent of donations coming from the general business community — large and small, agribusiness generally and families in rural and metropolitan Australia.

The Fund remains a potent force with selective financial support made available by an independent Board of Trustees, mainly to take actions likely to set a national legal precedent or which is of a broader community benefit.

The NFF is increasingly focusing its efforts on a range of issues affecting both the profitability and quality of life in the bush, including the environment and land use. Rural people are demanding equity with their city counterparts and NFF is making sure their voice is heard.

Lifelong learning is vital to agriculture's quest for improved productivity in an increasingly high technology future. People may change careers up to 5–6 times in their lifetime, with access to learning and skills' acquisition at an affordable price a crucial issue.

The NFF has campaigned vigorously on the issue of falling rural health service standards and achieved a major victory in 2000 with significant new federal government funding commitments to improve health services.

Gaining increased access to modern telecommunications technology has been another platform for the NFF, which has pushed hard for rural service upgrades and telecommunication facilities, which will enable them to participate in the global economy. NFF's communications subsidiary, Farmwide, has been a leader in offering country people access to internet services which can revolutionise the way they do business.

NFF has embraced the cause of sustainable farming, recognising the need for farmers to work in harmony with nature in a fragile landscape. The NFF has long been a strong supporter of the Landcare movement, a grassroots program, which addresses problems at the local community level.

The continuing use of precious natural resources, including water, is coming under increased challenge and scrutiny. We need to make sure governments are aware of the potential negative impacts on agriculture and the far reaching consequences of any decisions on natural resource allocation across the whole rural community.

ABOVE: NFF has embraced the cause of sustainable farming, recognising the need for farmers to work in harmony with nature in a fragile landscape.

SUSTAINING A NATION

NSW Agriculture

PROVIDING VITAL SUPPORT FOR RURAL COMMUNITIES

ABN: 86 308 026 589

161 Kite Street
Orange NSW 2800
Locked Bag 21
Orange NSW 2800

Phone: 02 6391 3100
Fax: 02 6391 3336
Email: nsw.agriculture@agric.nsw.gov.au
Website: www.agric.nsw.gov.au/

Dr Kevin Patrick Sheridan AO
Director-General of Agriculture

Dr Richard Frederick Sheldrake
Deputy Director-General of Agriculture

BELOW: Providing vital support to rural communities for 110 years.

FOR MORE THAN 110 years NSW Agriculture has provided vital support to the rural communities of New South Wales through its world-class research, extension, education and regulatory activities. This support has helped the state's primary producers achieve sustainable production of the highest quality, fresh, wholesome food and fibre products.

NSW Agriculture is committed to serving the rural sector in Australia's premier production state which contributes nearly 30 per cent of Australia's $28 billion farm income. Advice provided by extension officers helps producers to improve farm viability while maintaining a strong focus on sustaining the natural resource base.

NSW Agriculture's head office is located in Orange, in the central west of New South Wales. Nine centres of excellence and more than 80 regional offices are spread across the state, close to its 57,300-strong client base.

NSW Agriculture also supports joint community/departmental services such as Farming for the Future, the Rural Women's Network and Rural Lands Protection Boards.

The department also has a strong education focus. Two agricultural colleges provide opportunities for both young and mature-age students. The emphasis is on practical farm skills training to meet employment needs in a rapidly changing agricultural sector. For every job on a farm, four other jobs are created.

Agsell — NSW Agriculture's special export marketing assistance program — has made significant progress developing new business opportunities for producers. Excellent trade results have been achieved in China, Hong Kong, Korea, Vietnam and Japan in particular.

NSW Agriculture works in close harmony with producers, rural communities, agribusiness and government to achieve its vision of profitable, sustainable food

SUSTAINING A NATION

and fibre industries by constantly developing new technologies that significantly reduce input costs and environmental impacts, while boosting yields and quality.

Whole farm planning and monitoring systems are a popular initiative that deliver pasture and grazing systems expertise to nearly 1000 producer groups across the state.

An Irrigation Efficiency Unit advises farmers and horticulturists how to increase the productivity of high value crops, while reducing water use and preventing environmental damage to the soil. A Farm Forestry Advisory Unit helps producers to determine how farm forestry might best fit into their business plans.

Another valuable service to rural and urban communities is provided by grants of over $6 million a year to assist local government authorities with noxious weed control programs. NSW Agriculture's very active biological control program for weeds targets almost 50 weeds with over 30 established biological control agents, further reducing the reliance on herbicides.

Another initiative shows that early-season cotton insecticide use can be reduced by up to 47 per cent through adoption of new varieties and other technology. The state's pome fruit industry now uses 25 per cent less pesticide than in previous years and there has been a massive 90 per cent reduction in pesticide use to control banana weevil borer.

Other initiatives include 'one-stop-shops' for information on agriculture and a modern internet site that contains many thousands of pages of information on agricultural topics. Close links and working relationships with other state government agencies, universities, CSIRO, research and development corporations and industry, ensure that NSW Agriculture continues to make maximum use of scarce resources and facilities.

TOP: NSW Agriculture provides practical farm management advice that reduces environmental impacts while boosting yields and quality.

ABOVE: Research, education and extension is helping production of high quality, fresh, food and fibre products.

SUSTAINING A NATION

NSW Forest Products Association Ltd.

ECOLOGICALLY SUSTAINABLE FOREST MANAGEMENT

ABN: 96 001 866 468

Level 2, 60 York Street
Sydney NSW 2000
PO Box Q953
QVB NSW 1230

Phone: 02 9279 2344
Fax: 02 9279 2355
Email: forprod@ozemail.com.au

Mrs Lexie Hurford
President

Mr Colin Dorber
Executive Director

BELOW: Clearing land around the time of Federation.

THE TIMBER INDUSTRY in New South Wales has entered an era of sustained growth and development after decades of traumatic upheaval.

NSW Forest Products Association Executive Director Colin Dorber said there were many exciting opportunities now being taken to rebuild a strong rural forest industry based economy.

'Timber producers have shown their natural resilience during an extremely challenging period. Already, since the momentous outcomes of November 1998, over $100 million has been invested in technology and new equipment in the hardwood industry sector alone.'

The Forest Products Association, the forest industries peak representative body in New South Wales since 1906, believes recognition of the contribution forest products makes to the Australian economy and to society at large, will continue to emerge if the industry maintains its commitment to ecologically sustainable forest management and promotes its position, using science and sound economic facts to support its case.

Timber production is extremely important to the state economy with a gross output estimated at about $2.25 billion a year. Over 20,000 people are employed throughout the state, from growing to value added manufacturing and retailing.

The industry had been adjusting to reduced supplies of public forest resource by sustainably increasing harvests from private forest and plantations. Changing to products and markets that yielded higher returns with greatly improved utilisation of the forest resource will also enhance the growth prospects for the industry.

'Natural forests of hardwood, cypress and red gum provide the resource for several regional economies and are increasingly supplemented by hardwood and

SUSTAINING A NATION

exotic softwood plantations. This renewable resource is generating more highly skilled jobs and creating additional wealth for the community,' Colin Dorber said.

Changes to forest management and the allocation of the native forest resource reflect the increasing dependence upon regrowth and plantation timbers, which have very different qualities to the traditional resource base. Mills had responded by finding new markets, increasing value adding, expanding the scale of operations, improving recovery rates and investing in new plant and equipment.

Colin Dorber said the hardwood, cypress and red gum sectors has identified exciting development opportunities including the production and export of high value products featuring the natural beauty and strength of Australia's unique timbers.

Softwood plantations, whose product is focused at the high volume construction sector, would continue to provide a large and increasing share of sawn timber production, expected to reach about 70 per cent of total consumption this year.

As a result of the six-year state and federal government Regional Forest Aggreements process, the industry in New South Wales has moved to an ecologically sustainable resource base drawn from forests not required for conservation purposes, regrowth natural forests and hardwood and softwood plantations. New state legislation has led to a rapid increase in plantation area which has cut administrative red tape and boosted regional jobs.

Colin Dorber said there were a range of value-adding opportunities in export markets, particularly in North America, Europe and the Pacific Rim capitalising on the unique attributes of New South Wales hardwoods, cypress and redgum. For those committed to the principles of maximising utilisation, incorporating the benefits of carbon credits, ecologically sustainable forest management and maximised value adding at every level of the forest industry chain, there is a bright and positive future.

ABOVE: Timber getters with style. Photo of a water powered mill at Rockton, near Bombala, Southern New South Wales 1899.

BRIDGING THE FARMER–CUSTOMER GAP BRINGS MARKET GROWTH

ABN: 97 125 303 081

Royal Exchange Building
Level 7, 56 Pitt Street
Sydney NSW 2000
GPO Box 4806
Sydney NSW 2001

Phone: 02 9252 1344
Fax: 02 9252 2844
Email: nswgb-sydney@nswgb.com.au
Website: www.nswgb.com.au

Mr Don Hubbard
Chairman

Mr Graham Lawrence
Managing Director

BELOW: Topped and tamped bags of wheat at the end of the harvest.

A DEDICATION TO GRAIN GROWERS, quality, and attention to customer needs has been the force behind the growth of New South Wales' grain industry and has seen the NSW Grains Board (NSWGB) extend across domestic and international markets to achieve annual sales of $800 million a year with a record 3.5 million tonnes of accumulated grain.

In nine years the NSWGB increased its business by an average of 32 per cent per annum, with coarse grain and oilseed production in NSW achieving 50 per cent and 2300 per cent increases respectively.

The NSWGB has expanded along the eastern seaboard to provide marketing and supply services in New South Wales, Queensland, Victoria and South Australia. 'Our presence within eastern Australia's grain belt provides access to high quality product, a large storage network and builds strong relationships between growers and NSWGB grain marketers.'

The reputation of the NSWGB as a reliable supplier of clean, quality product enhances opportunities in the market and enables the NSWGB to maximise returns to growers.

The NSWGB takes in product through a network of some 500 country receival points with major inland grain terminals and export shipping facilities at Newcastle, Port Kembla, Brisbane, Geelong, Portland, and SA Ports, supplied by world-class grain storage companies.

The NSWGB followed on after 20 years of service from its forebears and was born in August 1991 from of the amalgamation of the four boards previously responsible for the marketing of barley, oats, oilseeds and grain sorghum. This amalgamation created a stronger single statutory authority responsible for marketing coarse grains and oilseeds in New South Wales.

The NSWGB existence preserves the time honoured and successful principle

SUSTAINING A NATION

of farmers controlling their own destiny in the marketing of their produce.

A continuous research effort makes market development for new grain varieties, end user specifications, improved growing methods, insect control, quality harvesting, handling and storage techniques a priority.

The NSWGB promotes and endorses best management practices at all stages of the grain supply chain from plant breeding to customer delivery.

Grower promotions, including the annual Good Oil Canola and Better Malting Quality awards, encourage farmers to strive for the quality expectations of customers. Seed quality analysis and constant testing from harvest to delivery give accurate indications of grain quality and product yield.

From a core business built on feed and malting barley, the NSWGB has diversified to handle significant tonnages of oilseeds, sorghum, pulses, domestic market wheat and other crops.

Much of its recent growth has come from oilseeds, providing the domestic grain industry with strong markets to match a rapid expansion in canola plantings, as well as sunflower, safflower and soybean production.

The NSWGB Beijing regional office is a key link to growth in the China market. Japan is another success story with close business links with major food conglomerates and trading houses in that country.

'Our focus on quality and on expanding to meet the needs of growers and customers has made the NSWGB Australia's most dynamic grain marketing organisation, and has formed a strong foundation to build on for the future.'

TOP LEFT: Golden Canola crop in southern New South Wales.

TOP: Barley harvesting, Gunnedah District, New South Wales.

ABOVE: End products of NSWGB marketed commodities.

SUSTAINING A NATION

QRAA
FINANCIAL SOLUTIONS FOR THE LAND

BUILDING VIABLE RURAL INDUSTRY IN QUEENSLAND

ABN: 30 644 268 943

Level 9, 307 Queen Street
Brisbane QLD 4000
GPO Box 211
Brisbane QLD 4001

Phone: 07 3370 0120
Fax: 07 3370 0180
Email: contact_us@qraa.qld.gov.au
Website: www.qraa.qld.gov.au

Mr Tony Ford
Chief Executive Officer

Mr Colin Holden
General Manager

BELOW: Loading sewn bags of wheat to deliver to silo depot.

QUEENSLAND RURAL ADJUSTMENT AUTHORITY, which operates under the Queensland Department of Primary Industries, is responsible for dispensing a mix of federal and state funds across the rural sector. This statutory authority has operated as a stand-alone entity since 1994 and works closely with banks and financial institutions to secure the future for producers. Rural producers in Queensland grappling with the pace of change and the challenges of building viable enterprises can call on this unique support agency and more than 5000 farmers a year do.

Chief Executive Officer Tony Ford said traditional support for farmers in times of natural disaster such as floods, cyclones or exceptional drought was still a core part of their business. 'But we're also very pro-active, supporting a range of innovative schemes, including FarmBiS, which look to the future.

'The authority has gained the respect of farmers across the state's diverse rural industries. We have a competent board that regularly meets in country areas and visits rural properties to gain first hand experience, whilst being supported by our team of strong, responsive and professional staff' Tony Ford said.

Practical programs such as the Primary Industry Productivity Enhancement Scheme attract strong interest.

Concessional loans are provided to assist in areas such as building property size, diversifying or acquiring a first property.

The authority has a strong Landcare focus, helping producers to rehabilitate and improve their properties to enhance long-term sustainability. 'We also address the important areas of training and professional development,' Tony Ford said.

SUSTAINING A NATION

THE LARGEST PRODUCER AND MARKETER OF STOCKFEED IN AUSTRALIA

RIDLEY AGRIPRODUCTS, the largest producer and marketer of stockfeed in Australia, is best known for brands including Barastoc, Supastok, Rumevite and Cobber.

The company has built its market leadership on scientifically formulated feed products backed by technical support to maximise the nutritional benefits of these products.

Ridley AgriProducts operates a network of 20 stockfeed mills and two supplement block plants throughout Australia. Altogether these plants have an annual production in excess of 1.3 million tonnes of high performance stockfeed. This feed services the needs of Australia's dairy, poultry, pig, beef, sheep and aquaculture industries. The company is also a leading producer of specialised feed for dogs, horses and laboratory animals.

Ridley AgriProducts is a division of Australian owned Ridley Corporation, a rapidly growing global business which is now the world's sixth largest stockfeed producer.

ABOVE: Ridley AgriProducts, state-of-the-art feedmill — Terang, Victoria.

ABN: 94 006 544 145

70–80 Bald Hill Road
Pakenham VIC 3810
PO Box 18
Pakenham VIC 3810

Phone: 03 5941 1633
Fax: 03 5941 3459
Email: enquiries@ridley.com.au
Website: www.agriproducts.com.au

Dr John S Keniry
Chairman

Mr John C Spragg
General Manager

BELOW: Ploughing the land about the time of Federation.

SUSTAINING A NATION

Ricegrowers
RICEGROWERS' CO-OPERATIVE LIMITED

FROM THE FARM TO THE INTERNATIONAL MARKET PLACE

ABN: 55 007 481 156

Yanco Avenue
Leeton NSW 2705
PO Box 561
Leeton NSW 2705

Phone: 02 6953 0411
Fax: 02 6953 4733
Email: mhedditch@ricegrowers.com.au
Website: www.ricegrowers.com.au

Mr Terry Hogan
Chairman

Mr Gary Helou
Chief Executive Officer

BELOW: Rice field in Leeton, NSW.

FROM HUMBLE ORIGINS 75 years ago, the New South Wales rice industry has emerged as a vertically integrated agribusiness of world class.

More than 2000 rice growing families are in control of their own destiny from the farm to the international marketplace through the successful operation of Ricegrowers Co-operative Limited.

The industry produces around 1.4 million tonnes of high quality paddy rice each year, almost entirely grown in the New South Wales Riverina irrigation areas. Ricegrowers receive, store, process and market the entire crop generating sales revenue around $700 million annually and providing thousands of jobs in southern regional New South Wales.

Australian rice growing got off to an uncertain start in the Murrumbidgee Valley in the early 1920s with growers struggling to survive the market domination of private milling companies. Fed up with continuing low returns, growers banded together to form Ricegrowers Co-operative in July 1950 and raised sufficient capital to build a rice mill at Leeton in time for the following harvest.

The new venture was so successful that Ricegrowers soon became the major processor and marketer of Australian rice. Today, Ricegrowers operates food processing plants at Leeton, Coleambally, Deniliquin and Echuca as well as modern stockfeed plants at Leeton and Tongala with a by-product conversion facility at Griffith.

The industry is market driven, producing a range of varieties sought by the marketplace for their quality. These range from the short grain Japonica varieties favored in the Japanese market to the long grain fragrant Jasmine rice well suited to all types of Asian cuisine.

New South Wales ricegrowers produce the highest yielding rice crops in the world averaging more than 9 tonnes a hectare. Yet farming is conducted in a

SUSTAINING A NATION

carefully structured way with an emphasis on environmental sustainability. The industry sets the benchmark for world's best practice with production restricted to heavy clay soils which limits groundwater recharge and results in high water use efficiency.

Ricegrowers has experienced a decade of aggressive growth and value creation in the run up to the new millennium with production doubling and revenue quadrupling.

About 85 per cent of Ricegrowers' products are sold into overseas markets making Ricegrowers Australia's largest exporter of branded food products.

Australian rice is now found in 72 countries with Papua New Guinea, Japan, Hong Kong and the Middle East among the biggest customers. Sales to Japan and the Middle East have risen rapidly in recent years. Offshore facilities are located in Papua New Guinea, Japan, Singapore and Dubai. Joint venture companies operate in England and Belgium.

Ricegrowers is famous for its SunRice brands which have been a marketing flagship since Sunwhite rice was introduced to the retail trade in 1955. The stable of SunRice brands includes a range of short, medium and long grain rices. Sunrice premium products include Jasmine fragrant, Mediterranean Arborio, Indian Basmati and Koshihikari Japanese rices marketed as 'SunRice Rices of the World' and the new Clever Rice with unique cooking qualities.

A range of value-added rice foods is also manufactured including SunRice rice cakes and a new range of innovative convenience rice foods including SunRice 3 Minute Rice to compete with the pasta and noodle sector and quick cooking Express Rice to suit busy lifestyles.

TOP LEFT: Australia has the highest yields of quality rice in the world, produced in an environmentally sustainable farming system.

TOP: A versatile, healthy and nutritious food suitable for the cuisine of any region, country, menu and occasion.

ABOVE: SunRice, the flagship brand of Ricegrowers' Co-operative, includes a range of high quality table rice products.

SUSTAINING A NATION 173

RURAL PRESS LIMITED

ONE OF THE WORLD'S LEADING REGIONAL AND AGRICULTURAL COMMUNICATIONS GROUPS

ABN: 47 000 010 382

159 Bells Line of Road
North Richmond NSW 2754
PO Box 999
North Richmond NSW 2754

Phone: 02 4570 4444
Fax: 02 4570 4663
Website: www.ruralpress.com

Mr John B Fairfax AM
Chairman of Directors

Mr Brian K McCarthy
Managing Director

BELOW: The road train has replaced the need to drive sheep along the road.

FROM A MODEST BEGINNING as the voice of country New South Wales almost 90 years ago, Rural Press Limited has grown into one of the world's leading regional and agricultural communications groups.

The company's roots are deeply embedded in rural Australia where the flagship title *The Land* was launched early last century. The specialist agricultural newspaper was born out of necessity to promote the interests of the fledgling NSW Farmers and Settlers Association.

Today, Rural Press Limited operates more than 210 agricultural and regional newspapers and magazines throughout Australia, the United States and New Zealand. It has an extensive web printing network, a stable of regional radio stations and is positioning itself to become a global force in regional electronic technology.

The company formed by farmers contributing six pounds each to an initial capital raising had a turnover last year exceeding $430 million on assets approaching $700 million.

Rural Press Managing Director Brian McCarthy said the modern company and its 3100 employees held to the same ethos of service to rural and regional communities spelt out in the first Editorial in *The Land* published on 27 January 1911.

'That Editorial made the point that a strong rural publication was wanted. People living in country areas needed an advocate to promote their views and stand up for their interests. That sentiment is as relevant today as ever in the regional markets in which we operate be they in Australia, New Zealand or the United States. The form of communications may be evolving as we cross the threshold of an electronic revolution, but our reason for being has not changed,' Mr McCarthy said.

SUSTAINING A NATION

For more than half its existence Rural Press remained a single title entity, building *The Land* from humble beginnings into one of Australia's foremost agricultural weeklies serving the most populous farming state. The modern era that was to transform rural publishing nationally and internationally began in 1975 with the purchase of competitor publication, the Sydney-based *New South Wales Country Life*.

The first move interstate came three years later with the acquisition of *Queensland Country Life*, a respected and influential weekly title serving rural producers in northern Australia.

Rural Press went national in 1980 with the launch of *FARM* magazine, the forerunner to the professional agribusiness title *Australian Farm Journal*.

During the 1980s Rural Press continued to build its strategic agricultural publishing interests across mainland Australia, with the acquisition of *Stock and Land* in Victoria, *Stock Journal* in South Australia and *Western Farmer & Grazier* in Western Australia, which merged into the publication known as *Farm Weekly*.

The beginnings of a regional publishing division, which now includes 160 titles, came in 1981 with the purchase of the northern NSW publications the *Armidale Express*, *Inverell Times* and *Glen Innes Examiner*.

This expansion has continued over the years in all states of Australia and the regional publishing division now encompasses such publications as the *Launceston Examiner*, *Ballarat Courier* and the company's only metropolitan daily *The Canberra Times*.

ABOVE: These days accessing virtual markets is becoming as important as weather forecasts and the Internet is a useful tool on the farm.

SUSTAINING A NATION

ABOVE: Rural Press has grown into one of the world's leading regional and agricultural communications groups operating more than 210 agricultural and regional newspapers and magazines throughout Australia, New Zealand and the United States.

The company's Broadcasting Division today holds six radio licences in South Australia and operates an FM station at Ipswich in Queensland.

The move to the international arena came across the Tasman in 1988 with the purchase of communication associates in Christchurch. Further acquisitions led to ownership of the National title, *New Zealand Farmer*, one of six specialist rural publications now part of the Rural Press group in that country.

The first move to expand into the United States came in 1989 with a series of acquisitions leading to the foundation of Rural Press USA. From its US headquarters in Carol Stream Illinois, the group operates 36 farm titles that serve more than 800,000 farmers across 48 states. The North American enterprise includes the operation of eight major farm shows across the US.

In Australia, Rural Press took over the running of the biggest agricultural event in the country, the annual AgQuip Field Days at Gunnedah in NSW through the purchase of BAL Marketing in 1995.

Through its purchase of regional publishing enterprises around Australia, Rural Press has put together the nation's most extensive commercial web printing network based at 18 sites across all states.

A milestone for the Printing Division came with the February 2000 opening of a state-of the-art $24 million facility at Rural Press headquarters at North Richmond, New South Wales. The culmination of five years planning, work began on the ambitious project in November 1998. The new press has a capacity of 70,000 copies an hour and can produce 192 tabloid pages including 48 pages in full colour in one pass, with in-line trimming, stitching and inserting.

In September 1998 Rural Press completed its biggest and most ambitious acquisition with the purchase of The Federal Capital Press of Australia, publishers of *The Canberra Times*. This development capped more than two decades of expansion in rural and regional Australia with the production of the daily newspaper in the nation's capital.

Rural Press has embraced the electronic information era with the launch of F@RMING OnLine, a gateway to the world for all things agricultural.

SUSTAINING A NATION

By combining its resources around Australia and overseas, the group has put together a website with an unparalleled collection of information and services.

The site includes up to the minute news and essential weather information, a calendar of major rural events, Australia's biggest virtual rural bookshop and sections devoted to employment, real estate and buy-sell classifieds.

The site also hosts forums for special interest groups across the farming spectrum.

A major Internet initiative was unveiled in July 2000 when Rural Press announced it had joined forces with Wesfarmers Dalgety Ltd and international strategy consultants McKinsey & Company to create a one-stop Internet portal for agriculture.

A $40 million capital raising finalised late in 2000, will bring in other strategic partners to help fund the development and operation of the ambitious Internet project.

The site will foster the development of rural e-commerce in Australia, based on the experience in the United States and has been timed to coincide with the rapid spread of reliable, fast Internet access across regional Australia.

Another exciting new development on the publishing front was announced in July when Rural Press unveiled plans to produce a new quality *Friday Magazine*.

The monthly magazine launched in August is inserted in key agricultural titles around Australia. High profile columnists contributing to the new publication include leading athlete Melinda Gainsford-Taylor, celebrity television chef Peter Howard and television journalist Ray Martin.

In June 2000 Rural Press moved into the outdoor advertising industry with its purchase of a 50 per cent stake in Street Vision, a company installing audio visual programming in Sydney's 10 underground railway stations. The technology is first to market, worldwide, and has applicability in both domestic and offshore underground railway systems.

ABOVE: The Land was first published on 27 January 1911. The first editorial made the point that people living in country areas needed an advocate to promote their views. That sentiment is as relevant today as it was then.

SUSTAINING A NATION

Sugar Research and Development Corporation

RESEARCH AND DEVELOPMENT EXCELLENCE

ABN: 41 343 997 980

Level 16, 141 Queen Street
Brisbane QLD 4000

Phone: 07 3210 0495
Fax: 07 3210 0506
Email: srdc@srdc.gov.au
Website: www.srdc.gov.au

Mr C P Hildebrand
Chairman

Mr E S Wallis
Executive Director

BELOW: Cane cutters before the era of mechanical harvesting.

THE BILLION DOLLAR export success enjoyed by the Australian sugar industry is built on a foundation of research and development excellence. The industry's research effort has, since 1990, been coordinated by a Commonwealth statutory authority, the Sugar Research and Development Corporation. A rapid expansion in sugar production during the last decade to a peak of 5.7 million tonnes in 1997 has been accompanied by a substantial increase in research funded by industry levies and matching Commonwealth Government contributions.

SRDC Executive Director Eoin Wallis said the Corporation's role was to direct research funding to foster an internationally competitive and sustainable Australian sugar industry.

The corporation has a five-year plan focused on improving productivity and cost efficiency of canegrowing and sugar manufacture. Among the key outcomes sought are enhanced marketability of Australian sugar and the sustainable use of land and water resources.

Eoin Wallis said that in addition to its core research funding, the corporation was directing $13.45 million of special Commonwealth funding into a four-year multi-disciplinary program to lift profitability in northern sugar regions.

Around 100 new projects are considered for funding each year with all research subject to performance evaluation and cost-benefit analysis.

'We have a selection process in place which aims to make the best possible use of the research dollar,' Eoin Wallis said.

While the industry has dedicated research bodies, the Bureau of Sugar Experiment Stations and the Sugar Research Institute, increased funding is now being directed to groups such as universities, CSIRO and state government departments. Collaborative research across agencies is also encouraged.

SUSTAINING A NATION

EFFECTIVE REPRESENTATION FOR CANE GROWERS

AT A TIME when the sugar industry is enduring one of the most challenging periods in its history, farmers are benefiting from effective representation by their organisation, CANEGROWERS.

A combination of collapsed world prices, poor seasons and major structural and legislative changes have impacted upon the future for Queensland's 6500 cane growing families. CANEGROWERS General Manager Ian Ballantyne said substantial benefits had been achieved for members under very difficult circumstances.

The growers own organisation had not been immune to change with a restructure into a public company in January 2000 after sweeping reforms to government legislation which dated back to the 1920s.

'Our organisation is entirely funded by growers and we are focused on providing the best possible representation and services to secure their profitable future,' Ian Ballantyne said. 'However, even in these tough times, our financial membership will be in the order of 95 per cent of all cane growers.'

CANEGROWERS was committed to highlighting the dire financial plight facing sugar producers and played a leading role in efforts to gain government assistance.

'Farmers in the far north have been hit particularly hard with severe cyclone and flood damage early in 2000 exacerbating their financial situation.'

CANEGROWERS has also been active in reviewing major changes to industry legislation which devolves much of the regulatory structure to agreements at the local sugar mill level. Extensive negotiations have been held to protect the interests of growers and their rights to collective bargaining.

The growers' organisation was also heavily involved in the restructure of sugar marketing and the transfer of title of Queensland's seven bulk sugar terminals to the industry.

SUSTAINING A NATION

ABN: 94 089 992 969

190–194 Edward Street
Brisbane QLD 4000
GPO Box 1032
Brisbane QLD 4001

Phone: 07 3864 6444
Fax: 07 3864 6429
Email: canegrowers@canegrowers.com.au
Website: www.canegrowers.com.au

Mr Harry Bonanno
Chairman

Mr Ian Ballantyne
General Manager

BELOW: Bob Ritchie with employer Alan Cole in Queensland in the 1970s.

AUSTRALIA'S SEAFOOD CENTRE OF EXCELLENCE

ACN: 064 254 306

Bank Street
Pyrmont NSW 2009
Locked Bag 247
Pyrmont NSW 2009

Phone: 02 9660 1611
Fax: 02 9552 3632
Website:
www.sydneyfishmarket.com.au

Mr Bill Gibson
Chairman

Mr Grahame Turk
Chief Executive

BELOW: The most famous of all our seafoods is undoubtedly the prawn.

SYDNEY FISH MARKET, a world-class food marketing complex that seamlessly links the value chain between producers and consumers, is to be found just a stone's throw from Sydney's central business district.

Embracing the challenge of industry deregulation, Sydney Fish Market has successfully positioned itself as Australia's seafood centre of excellence. The thriving commercial hub combines the biggest fish-trading auction in the Southern Hemisphere with quality retail and dining facilities.

The result is a waterfront facility at Pyrmont which has become a magnet, drawing almost two million shoppers, visitors and tourists a year. Every weekday the Sydney Fish Market auction offers more than 100 species of fish and crustacea from around Australia and overseas. For sheer diversity it is second only to the famous Tsukiji Market in Tokyo, Japan.

Sydney Fish Market Chief Executive Grahame Turk said the Market had ensured its relevance into the future by adopting a whole of industry approach.

'It is our business to service the needs of fishermen, wholesalers, retailers, restauranteurs and consumers,' Grahame said.

Since it was privatised in 1994, the Sydney Fish Market has operated as a joint venture company, 50 per cent owned by NSW fishermen and 50 per cent by tenants on the site.

Modern technology has been introduced into the sale system where more than 70 tonnes of product changes hands on a typical trading day. The Market has adapted the unique Dutch auction system, devised originally to sell flowers efficiently in Holland.

Every day about 170 buyers bid via computer terminals which fully automate the sale process. About 90 per cent of product is sold in this fashion. Traditional voice auction has been retained for live product such as mud crabs

SUSTAINING A NATION

and lobster, while there is a special buying pavilion for high-grade sashimi. The company is also currently developing an e-commerce facility which will enable seafood to be traded online by early 2001.

Six of Australia's largest seafood retailers operate on site and there is a selection of restaurants, plus a sushi bar, bottle shop, delicatessen, bakery, and shops selling fruit and vegetables, oysters, net and tackle, as well as poultry and gifts.

The Sydney Fish Market plays a pivotal role in boosting seafood consumption with regular seasonal promotions and the production of promotional material, recipes and consumer information made available to the public. The nutritional benefits of eating fish are highlighted through brochures, Fish Line — a consumer information line, and a website.

The Market is also home to the Sydney Seafood School, a purpose-built teaching facility designed to educate the general public and seafood chefs in all aspects of seafood handling and cooking. The school has been successful in promoting the knowledge and cooking skills for the many diverse species of seafood.

In an effort to set a seafood industry benchmark for quality, Sydney Fish Market Pty Limited has been certified to ISO 9002 and HACCP. Implementing and maintaining this certification ensures that consumers consistently receive only the highest quality, fresh seafood.

The Market also encourages industry excellence across all sectors by hosting the Sydney Fish Market Seafood Awards every two years.

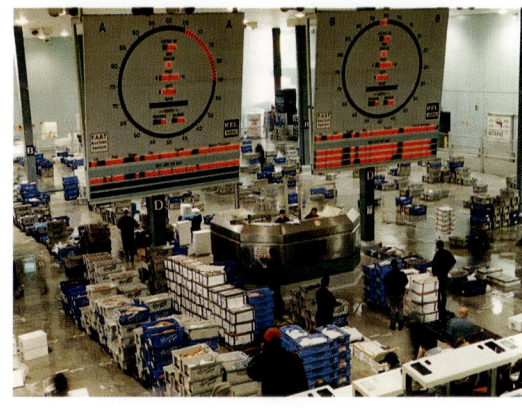

TOP: Alfresco dining area.

ABOVE: Sydney Fish Market auction floor.

SUSTAINING A NATION

SYDNEY MARKETS LIMITED

THE COMMERCIAL HEART OF THE AUSTRALIAN HORTICULTURAL INDUSTRY

ABN: 51 077 119 290

Level 3, Market Plaza Building
Sydney Markets NSW 2129
PO Box 2
Sydney Markets NSW 2129

Phone: 02 9325 6200
Fax: 02 9325 6288
Email: info@sydneymarkets.com.au
Website: www.sydneymarkets.com.au

Mr Geoff Bell
Chief Executive Officer

Mr Eric Kime
Chairman

BELOW: Markets have been part of Sydney agriculture since early settlement.

WITH A THROUGHPUT of one million tonnes of fresh produce a year, Sydney Markets is one of the largest food distribution centres in the Southern Hemisphere, and combines a unique blend of wholesale and retail trading with fresh produce consigned daily from all over the country.

With sheer volume of sales worth about $1.5 billion annually, Sydney is the national barometer for market prices.

There is action 24 hours a day on the 41 ha main market site, which forms its own mini suburb. More than 5000 people work within the markets catering for the professional buyers as well as about 100,000 consumers a week seeking produce bargains.

From early evening an estimated 500 semi-trailers and farm trucks begin arriving and unloading for the next day's business. More than 700 forklifts operate on site shuttling produce onto trading stands, in and out of cold rooms and to buyers' transport for delivery.

The vast multi-cultural trading facility was purpose built in 1975 as a New South Wales government owned entity, which made the transition to a privatised company in 1998.

Ownership is now vested in the many business enterprises, which trade in the markets through various share categories. An experienced 10-member board comprises nominees of five different classes of shareholders plus independent directors under the chairmanship of qualified accountant Eric Kime.

Chief Executive Officer Geoff Bell said the success of privatisation was clearly demonstrated with a strong, viable corporate entity showing a healthy surplus in its third year of trading.

Geoff Bell said Sydney Markets had completed a wide range of new site works and was dedicated to providing the best environment to support

SUSTAINING A NATION

competitive trade and effective distribution of a myriad of products.

In addition to a huge daily domestic turnover, Sydney Markets is also a major hub for the export of fresh produce from Australia to all parts of the world.

The market sells produce from about 20,000 farms and typically draws about 600 retailers each trading day.

The major trading centre includes the premises of 150 wholesalers and more than 400 mostly local growers who trade their own produce.

The adjacent Sydney Flower Market is the only facility of its kind in Australia open to the public as well as traders.

More than 180 growers have registered stands offering more than 200 different varieties of flowers and foliage.

Saturdays in Sydney Markets are filled with thousands of consumers with more than 800 stands carrying a huge full range of fresh fruit and vegetables as well as all manner of produce from eggs and honey to seafood, nuts and breads.

A separate second-hand Swap and Sell Market which also operates on Saturdays is known as Australia's biggest garage sale.

Sydney's Paddy's Markets also attract flocks of bargain hunters trading at Homebush on Friday and Sunday and on a separate site at Haymarket on Thursday, Friday, Saturday and Sunday.

TOP LEFT: The Sydney Flower Market is Australia's largest flower market, with an annual turnover in excess of $100 million.

TOP: 100 million tonnes of fresh produce go through Sydney Markets each year.

ABOVE: Customers range from national supermarket chains to small independent grocers.

SUSTAINING A NATION

TWYNAM Agricultural Group

TOTAL COMMITMENT TO SUSTAINABLE RESOURCE DEVELOPMENT FOR REGIONAL AUSTRALIA

Level 7, 17 Bridge Street
Sydney NSW 2000

Phone: 02 9325 9000
Fax: 02 9325 9050
Email: mail@twynam.com
Website: www.twynam.com

Mr John Kahlbetzer
Chairman

Mrs Christine Campbell
Chief Executive Officer

BELOW: Wool exports have been 'sustaining the nation' for nearly 200 years.

THE TWYNAM AGRICULTURAL GROUP was commenced in the early 1970s by its Chairman John Dieter Kahlbetzer, founded upon his strong affinity with rural Australia. This affinity with agriculture, which is shared by his two sons Johnny and Markus, also led to the foundation of the Liag Group in Argentina in 1982.

Today, Liag is active in the production of irrigated and dryland cropping, cattle breeding and fattening. Liag has contributed to the introduction of new technologies and improved management practices, committing to the development and sustainability of the regional communities in Argentina.

During the 1970s, Twynam acquired a number of mixed farming properties throughout New South Wales including the part of the Amatil property aggregation that included the renowned Mungadal Stud at Hay, Jemalong at Forbes and Buttabone at Warren.

Throughout the 1980s and 1990s, Twynam saw the potential of the irrigation industry for efficient agricultural production and expanded its interests in the Murrumbidgee, Lachlan, Macquarie, Barwon and Gwydir river valleys.

Today, the Twynam and Liag property portfolios have a wide geographical base expanding over 540,000 hectares with a sector spread providing security against adverse weather conditions and commodity market fluctuations. The Group's rural industry mix now includes beef, cattle feedlotting, wool, rice, cotton and cereals. Twynam is one of the largest cotton, rice and cereal growers in Australia. With the acquisition of the Colly Cotton Group in 1999, Twynam now operates three cotton gins in New South Wales. Through its Growers Services division, Colly Cotton has extensive coverage of the cotton producing areas in Australia and markets about 20 per cent of Australia's cotton crop internationally.

SUSTAINING A NATION

The Chief Executive Officer, Christine Campbell, who has worked for the Group since the late 1970s, said the Kahlbetzer family were passionate about their rural activities. 'It is very much a people oriented business. There is a strong commitment to participate in a progressive and sustainable future for the regional areas of Australia,' she said.

Twynam supports programs for on-the-job training, offers rural cadet scholarships and encourages rural leadership programs to ensure competent leaders of the future for the industry. Staff adopt best management practice and are encouraged to participate in rural industry and regional programs sharing the knowledge acquired from trialling new techniques.

Twynam participates in research and development projects across all rural industries directing efforts into industry CRCs and the CSIRO in particular. The Group's sponsorship of the Chair of Animal Breeding Technologies at the University of New England will lead to the use of new technologies by all breeders at an affordable cost.

Christine Campbell states, 'as a vertically integrated participant in the agricultural industry, the Twynam Agricultural Group has a total commitment in ensuring sustainable resource development for the long term future of its Group, its employees, and for rural communities in partnership with our environment'.

TOP: Twynam's Mungadal Stud at Hay, Australia's 15th oldest Merino Stud.

TOP (INSET): Mr John Kahlbetzer, the Chairman of the Twynam Agricultural Group.

ABOVE: Twynam's 30,000 head cattle herd has significant Hereford bloodlines.

SUSTAINING A NATION

Food Victoria

State Government Victoria — Department of State and Regional Development

FOOD VICTORIA — THE FOOD SECTOR AND GOVERNMENT GROWING THE INDUSTRY TOGETHER

FOOD VICTORIA — Department of State and Regional Development

ABN: 69 981 308 782

GPO Box 4509RR
Melbourne VIC 3001

Phone: 03 9651 9444
Fax: 03 9651 9304
Email: food@food.vic.gov.au
Website: www.food.vic.gov.au

BELOW: Australia is building a world-wide reputation for 'clean green' food.

AGRICULTURAL PRODUCTION WORTH $6.1 billion per annum and processed food products with an ex factory value of more than $14 billion a year make agri-food Victoria's largest industry.

Exports of food and fibre products total $5.2 billion a year, and represent 31 per cent of the state's export earnings. Exports have more than doubled since 1992, and include markets as diverse as Japan and Switzerland, Singapore and Mexico.

More than $2.5 billion has been invested in food processing in Victoria since 1992, indicating a high level of confidence in the future of the industry. Approximately 42 per cent of food manufacturing jobs are in regional Victoria, which accounts for 21 per cent of rural employment. Many other jobs are directly linked to the food industry, including chemical suppliers, packaging and transport.

Food Victoria provides a whole-of-government umbrella under which a range of individual food sector activities are coordinated. These include infrastructure, agriculture, business development, investment, research, training and food safety.

The Food Industry Advisory Committee (FIAC) is chaired by the Premier and guides Food Victoria's priorities. FIAC currently comprises the Minister for State and Regional Development and the Minister for Agriculture and senior representatives of primary producers, large and small food processors, distributors, retailers, exporters, unions and consumer groups. This ensures that all stakeholders have direct input into policy development and regulation reform. Government action in response is coordinated through a Food Industry Taskforce, which is chaired by the Convenor of Food Victoria and comprises officials from across six agencies.

SUSTAINING A NATION

Natural Resources and Environment
AGRICULTURE • RESOURCES • CONSERVATION • LAND MANAGEMENT

This partnership between government and the industry has set a food and fibre export target of $12 billion by 2010. Its vision includes many components, including the view that growth is export driven, sustainability is crucial, efficiency gains must be maintained to ensure competitive and profitable businesses, continuing investment must be encouraged, and the sector must be supported by best practice in food safety, relevant world class research and development, and a skilled workforce at all levels.

Food Victoria initiated a joint venture between the Australian Food Industry Science Centre and the CSIRO, to create Food Science Australia, the premier food research organisation in Australia. A FIAC member is Chair. Food Victoria's communications and promotions strategy includes a website, newsletters, a CD-ROM showcase and many other publications, and support for major events such as food and wine festivals.

The current work program of Food Victoria has identified the key priorities to be water resource management and sustainability, research and development, and skills development, education and training.

Current government programs include the Naturally Victorian initiative to improve the marketing performances of farmers and small food producers, and the provision of substantially increased resources to support the development of rural and regional Victoria, particularly in regional industries such as food and forestry.

Food Victoria maintains strong links with Supermarket to Asia Ltd and Commonwealth agencies to provide informed input on key national issues such as overseas market access, single desk trading, food standards, food safety legislation, genetically modified foods, and demand chain management.

AGRIBUSINESS GROUP —
Department of Natural Resources and Environment

ABN: 90 719 052 204

Level 15, 8 Nicholson Street
East Melbourne VIC 3002

Phone: 03 9637 8085
Fax: 03 9637 8119
Website: www.nre.vic.gov.au

BELOW: Victoria has set a food/fibre export target of $12 billion by 2010.

SUSTAINING A NATION

Westpac Business Banking

BANKING PIONEER LOOKS TO THE FUTURE

ABN: 33 007 457 141

Level 24, 60 Martin Place
Sydney NSW 2000

Phone: 02 9226 6355
Fax: 02 9216 0665
Email: bbuffier@westpac.com.au
Website: www.westpac.com.au

Mr Barry Buffier
*General Manager NSW and
Agribusiness Regional Banking*

BELOW: 'Oberon' the bull has a savings account in which $10 is deposited each time he sires a calf.

WESTPAC has been part of the social fabric of the nation since the first branch was set up in the struggling New South Wales penal settlement in 1817. It is proud of its heritage as Australia's oldest existing bank and in its third century of operations remains as determined as ever to be the preferred financier for the rural sector.

'We've been there for all the important milestones, from the gold rushes to the era when our rural industry set the backbone of the new economy with the first shipments of frozen beef and significant wool exports,' says Graham Jennings, National Manager, Regional Banking. 'Our commitment to regional Australia spans more than 180 years and now Westpac has created a separate division of "Regional Banking" to manage all of our activities in Regional and Rural Australia' (RARA).

'Our commitment cannot be expressed any clearer than by the fact that for ever dollar of deposits we raise in RARA, we reinvest nearly two dollars in loans for RARA. This is no small sector. It represents 1.2 million of our customers who borrow $15 billion.

'As well as a structure which is firmly focused on RARA, we are using modern communication technology to even better position the bank to serve the needs of these customers.' says Graham.

'We are establishing Regional Business Banking Centres to service our rural customers and to support our mobile Agribusiness and Relationship Managers in the field.'

Regional Business Banking Centres will be established at Townsville, Orange, Ballarat, Adelaide and Perth. It is complemented by more than 40 Regional Finance Centres offering an extensive range of lending and investment advisory services in key provincial cities and towns.

SUSTAINING A NATION

Heading up the Agribusiness side of Regional Banking for Westpac is Barry Buffier, previously Deputy Director General of NSW Agriculture, who joined Westpac in 1993.

Barry says that Westpac is a committed lender to the rural sector as it represents a good business opportunity for the bank. RARA is a vital component of the Australian economy and Australian agriculture benchmarks with the best in the world.

'Australian producers are highly innovative and internationally competitive at the farm gate level without significant government subsidies. We can capitalise on our clean green image and we are well located for the all-important Asian market. Many agricultural industries have annual productivity growth of around 4 per cent, far exceeding their overseas competitors and many other industries in Australia,' says Barry.

Westpac has not just been a passive financier to the rural sector, but has been proactive in the national policy debate on issues affecting rural Australia.

'We took a leading role with NFF and state farmer organisations in the push to have Farm Management Deposits privatised because we realised their potential to help our customers deal with the cyclical nature of agriculture,' says Barry.

'We have had input to the debate surrounding significant environmental issues such as Australia's worsening salinity problem and we actively support the Landcare movement. Such a large part of our business depends on the health of the rural sector. It's in our interest and our customers' interest for Westpac to become involved, and help secure a long-term sustainable future for rural Australia.'

TOP: Lang diesel tractor and Massey Harris Reafer Musher on MacCallum Smiths' farm, Koora.

ABOVE: Bank of New South Wales Head Office, Wool Department, 1905.

SUSTAINING A NATION

www.focus.com.au

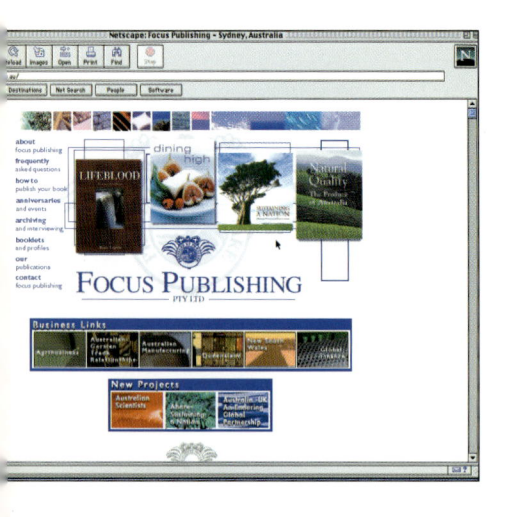

Visit the Focus website for information on Focus Publishing, Australia's leading corporate book publisher. Focus specialises in producing high-quality custom books, corporate histories and specific marketing, event, promotional and anniversary books.

To obtain further information on the companies and organisations participating in this book, *Sustaining a Nation*, simply follow the steps.

- Enter the Focus Website address:
  www.focus.com.au
- You are now on Focus' home page. Click on the words our publications.
- Scroll down to the bottom of the page and click on latest publications.
- Scroll down the page to find *Sustaining a Nation*.
- Click on Business Links. You will find the websites of the participating companies and other organisations listed in light blue type.
- Click on your chosen company's or organisation's address and it will lead you to their website.

ABARE	www.abareconomics.com	NSW Agriculture	www.agric.nsw.gov.au/
Agriculture, Fisheries and Forestry — Australia	www.affa.gov.au	NSW Grains Board	www.nswgb.com.au
Australian Fisheries Management Authority	www.afma.gov.au	Queensland Rural Adjustment Authority	www.qraa.qld.gov.au
Australian Seafood Industry Council	www.asic.org.au	Ridley AgriProducts Pty Ltd	www.agriproducts.com.au
Avcare Ltd	www.avcare.org.au	Ricegrowers' Co-operative Limited	www.ricegrowers.com.au
Aventis CropScience Pty Ltd	www.aventis.com.au	Rural Press Limited	www.ruralpress.com
CANEGROWERS	www.canegrowers.com.au	Sugar Research and Development Corporation	www.srdc.gov.au
Consolidated Meat Group Pty Ltd	www.meat.aust.com	Sydney Fish Market Pty Limited	www.sydneyfishmarket.com.au
Consolidated Pastoral Company Pty Ltd	www.pastoral.com	Sydney Markets Limited	www.sydneymarkets.com.au
FreightCorp	www.freightcorp.com.au	Twynam Agricultural Group	www.twynam.com
Goodman Fielder	www.goodmanfielder.com.au	Victorian Department of Natural Resources and Environment — Agribusiness Group	www.nre.vic.gov.au
GrainCorp	www.graincorp.com.au		
Manildra Flour Mills Pty Ltd	www.manildra.com.au	Victorian Department of State and Regional Development — Food Victoria	www.food.vic.gov.au
Murray Goulburn Co-operative Co Limited	www.devondale.com		
National Farmers' Federation	www.nff.org.au	Westpac Banking Corporation	www.westpac.com.au

INDEX

Page numbers in *italics* refer to illustrations.

abalone farming 73
ABARE (Australian Bureau of Agricultural and Resource Economics) 7, 135, 136–9, 190
Aborigines in pastoral industry 96
agricultural shows 103–4, *107*, *109*
agricultural society movement 103
Agriculture, Fisheries and Forestry — Australia 135, 140, 190
'Air Beef' 116
air transport *114*
angora wool 93
apples 65
aquaculture 73
artesian water *129*
Australian Bureau of Agricultural and Resource Economics *see* ABARE
Australian Fisheries Management Authority 135, 141, 190
Australian Seafood Industry Council 135, 142, 190
automation *see* mechanisation
Avcare Ltd 135, 143, 190
Aventis CropScience Pty Ltd 135, 144–5, 190

BAE (Bureau of Agricultural Economics) 7
barley 34, 36
beef 11–12, *18–19*
blowfly strike 92
bores, water *129*
bounty, sugar 43
branding *20*
brewing 36
Britain: dairy exports to 24; meat export agreement 11–12, 116; sugar export to 45; wartime trade agreement 23; wool export to 91
broadacre crops *32–41*, 33–36
bulk handling 116, *119*, *120–21*
bullock trains 115
Bureau of Agricultural Economics (BAE) 7
butter 23–24, *27*, 116

calicivirus 127
cane toad 127–28
cane cutters 43, *44*, *46*
CANEGROWERS 135, 179, 190
canning 62–64
canola (rapeseed) 36, *41*
carrots *63*
casks, wine *58*, *59*
cattle: beef *10*, 11–12, *12*, *18–19*, *20–21*; dairy 22, *28*; diseases 126; sales *112–13*
cereal crops *32–41*, 33–36
chaff *37*
cheese 23–24, 24–26, *26*
chickens *12*, *14*, *15*
chickpeas 36
child labor 29, *102*
chilling 115, *122*
Chinese immigrants 47, 61, 64, 66
cholesterol 26
citrus 62
clearance of land *18–19*, 128, *132*
clearfelling 83
closer settlement schemes 5–6, 23, 33, 43
clover 126
cold storage 115
commodity forecasts and projections 7
community life 5, 6, *102–12*, 103–6
competitions *107*
conservationists 83
Consolidated Meat Group Pty Ltd 135, 147, 190
Consolidated Pastoral Company Pty Ltd 135, 146, 190
consumption: butter 24; cheese 26; chicken *15*; fish 72; fruit and vegetables 64; meat 11, 12–13, *17*
container shipping 116
cooperative factory system 23, *27*
cotton 93, *98–99*
cottonseed 36
country towns 103–5
Country Women's Association (CWA) 104, *104*, *105*
cream separators *27*
crocodile meat 13
crops: cereal *32–41*, 33–36; horticulture *60–69*, 61–64; sugar *42–49*, 43–45
CSIRO 125
cultured pearl farms 71, *76*
custard 26
cutflowers *60*, 64

dairies *29*
dairying *22–31*, 23–26
Dairy Farmers 135, 148–51, 190
deepwater trawling 72
deer industry 93
Depression 6, 62, 91
deregulation: fresh milk market 24; wheat marketing 34
dietary concerns: dairy products 26; fruit and vegetables 64; meat 11, 12
dipping: cattle *12*; sheep 92
disease prevention 24, 126, *132*
dogs in sheepfarming 97
draught horses *108*
dried fruit industry 62, *62*
drought *130*
droving 115
duffing *20–21*
dust storms *131*

egg production *14*
emu meat 13
environment 7, 83
erosion 125
eucalypts *85*, *89*
European Economic Community 24
exports: apples 65; beef 11; dairy produce 23–24; flowers *60*; impact of improved transport 115; lamb 12; live sheep 12, *120*; modern trends 116–17; seafoods 71; sugar 44–45; timber 81; wheat 34; wine 51, 52; woodchips 83; wool 91

faba beans 36
fallowing 33
family farms 7, *28*
Federation 6
feedlot industry 20, 36
fertilisers 126
fibres *90–99*, 91–93
field days *106*
field peas 36
fisheries *70–78*, 71–73
fishing zone 73
flour 33
flowers *60*, 64
fodder crops *32*, 36, *37*
forestry *80–89*, 81–83
'free range' poultry *14*
freezing 115
FreightCorp 135, 152–3, 190
fruit growing 62, *62–64*, *65*, *68*

German immigrants 57
globalisation 7, 116
Goodman Fielder 135, 154–5, 190
GrainCorp 135, 156, 190
grains *32–41*, 33–36, *120–21*
grapes, wine *50–59*, 51–53
Greek immigrants 72
gross domestic product (GDP) 5
guaranteed home consumption price 34

handling systems 116
harvesters (machines) *35*, *39*
hay *32*, *37*
horses *108*
horticulture *60–69*, 61–64

immigrants: fisheries 71–72; horticulture 61, 64, 69; sugar industry 46; tobacco industry 66; wine industry 51, 52, 57
indentured labor 43, 46
internet 105, 117
irrigation *64*, 128
irrigation settlements 61–62
Italian immigrants 46, 51, 57, 66, 71

jam making 62–64
Japan, beef export to 12

Kanakas 43, *46*, *47*
kangaroo meat 13
kelpie *97*
knitting 96

lamb 12, *13*
Land Army girls *67*, *110*
land clearing *18–19*, 128, *132*
Landcare 133
lentils 36
linseed 36
live sheep export 12, *120*
lupin 36

maize 36
malting 36
Manildra Flour Mills Pty Ltd 135, 157, 190
margarine 24
market gardening 61, 64
marketing boards: grain crops 34, 36; sugar 44; vegetables 65; wool 92
meat *10–20*, 11–13; transportation 11, 116, *122*
mechanisation: canning 64; consequences 104; cotton harvesting *98*; dairy industry 24, *29*; forestry *87*, *88*; sheepshearing *94*, *107*; sugar industry 45, *48*; wheat growing 33, 34; wine industry 53, *56*, *57*; World War II 6–7
MIA Council of Horticultural Associations 135, 158
Middle East, beef export to 12
milk *22*, 24, *24–25*, 115
milk tankers *30–31*
milking machines 23, *29*
milling plant, sugar *49*
mother-of-pearl 71
Murray Goulbourn Co-operative Co. Limited 135, 159, 190
Murray River 51, 61, *64*, *74*
Murrumbidgee irrigation area 34, 51, 62
mustering *18*
mutton 11, 12
myxomatosis virus 127

National Farmers' Federation 135, 160–3, 190
national parks 82
native forest 83, *87*
NSW Agriculture 135, 164–5, 190
NSW Forest Products Association Limited 135, 166–7
NSW Grains Board 135, 168–9, 190

oats 36, *36*
oilseeds 36, *41*
opium poppies *67*

paper industry 89

SUSTAINING A NATION · 191

pasteurisation 23
pearl farms 71, 77
pearling 70, 71
pest control 126–8, 128
pigs 12, 17
plantation system (sugar industry) 43
pork 12, 17
potatoes 65
poultry 12, 14, 15
prawns 71
price support schemes 24, 116
prices: dairy produce 23–24; depression of 7; government-set 116; sugar 43–44; wheat 33–34; wool 91, 92
prickly pear 126, 127
prime lamb production 12 13
processed cheese 26
productivity growth 105
protective trade barriers 116
pulse crops 36
pumpkin 66

quarantine restrictions 126
Queensland Rural Adjustment Authority (QRAA) 135, 170, 190

rabbits: as food 16; as pests 126–27
railways 30, 33, 45, 47, 49, 115
rain forests 80
rapeseed (canola) 36, 41
refrigerated transport: beef and mutton trade 11; dairy products 23, 31; impact 115, 122
regulation 24, 116
regulatory bodies see marketing boards
research: CSIRO 125; veterinary 126; viticultural 53
rice 34, 38, 105
Ricegrowers' Co-operative Limited 135, 172–3, 190
Ridley AgriProducts 135, 171, 190
Riverina 34, 36, 38, 105
road trains 115, 123
rodeos 108
roller milling plant 49
rural life 5, 6, 102–12, 103–6
Rural Press Limited 135, 174–7, 190

safflower 36
salinity 64, 128, 130
sawmills 82, 84
school milk scheme 25
seafoods 71–73
seine fishing 72
selectors 6
settlement schemes: closer settlement schemes 5–6, 23, 33, 43; irrigation settlements 61–62, 64; soldier settlement schemes 6, 7, 62
shearing 94–95, 107
sheaves 36

sheep: diseases 126, 132; export 120; lamb and mutton 11, 12, 13; transportation 114, 117; wool 90, 91–93, 92–97
sheepdogs 97
shipping 86, 116, 120–21
sideshow alley 104, 109
silage 28
silos 116, 119
skeleton weed 126
soil conservation 125–26
soldier settlement schemes 6, 7, 62
sorghum 36
Spanish community 68
state forests 82
stockfeed 32, 36, 37
stooks 36
sugar 42–49, 43–45, 116
Sugar Research and Development Corporation 135, 178, 190
sultanas 62
sunflower seed 36
supermarkets 68–69
superphosphate 33, 126
sustainable agriculture 124–33, 125–28
synthetic fibres 92
Sydney Fish Market 78–9, 135, 180–1, 190
Sydney Markets Limited 135, 182–3, 190

tankers, milk 30–31

tariffs 51, 116
temperance movement 51, 54
'tick gates' 12
timber 80–89, 81–83
tobacco 66
tomatoes 69
towns 103–5
trace element deficiencies 126
transportation 114–23, 115–17; meat 11; milk 30–31; sugar 45, 45, 47, 49; timber 86
triticale 36
tuna 73
Twynam Agricultural Group 135, 184–5, 190

United States, exports to: beef 12, 18–19; sugar 45; wool 91
USSR, beef export to 12

vaccines 126
value-adding industries 105
vegetables 63, 64, 65–66, 68–69
veterinary science 126
Victorian Department of State and Regional Development — Food Victoria 135, 186–7, 190
Victorian Department of Natural Resources and Environment — Agribusiness Group 135, 186–7
'virtual markets' 117
viticulture 50–59, 51–53

water: conservation 128, 129; irrigation 61–62, 64
weeds 126
Westpac Banking Corporation 135, 188–9, 190
whaling 72, 75
wheat: handling 116, 118, 119; production 33–34, 40; transportation 120–21, 122
'white Australia' policy 43, 66
'white grown' sugar cane 43, 47
wild flowers 60, 64
wine 50–59, 51–53
wine bars 54–55
women in agriculture 67, 104, 110–11
Women's Land Army 67, 110
woodchipping 83, 89
wool 90, 91–93, 93–97; handling 118; sale methods 117; transportation 115
Woolmark logo 92
World War I: government regulation 116; rabbits 16; soldier settlement schemes 6; wheat production 33
World War II: dairy industry 24; horticulture 63; mechanisation 7; road construction 115; soldier settlement 7; wheat production 34; Women's Land Army 67, 110; wool exports 91

Yugoslav migrants 52, 71

PICTURE CREDITS

Focus Publishing would like to thank the many individuals and organisations who provided assistance in sourcing images for this book. The following photographs are reproduced courtesy —

ABARE: p17 (inset), pp18–19, p20 (bottom), p21, p22, p29 (top), p32, p39, p40, p41, p57, p73, pp74–5, p76 (top), p78, p79, p87, p89, p95, p97 (bottom), p98 (bottom), p99, p111, p120, p123, p124, p130, endpapers.

AFFA: p133.

Agriculture Western Australia: p60.

Australian Centre for Public History: p5, p25, p28 (bottom), p29 (bottom), p30, p34, p38, p44 (bottom), p52, p53, p54, p55, p56, p57 (inset), p58, p63, p66 (top), p68 (top), p72, p84 (top), p92, p96 (top), p98 (top), p110 (bottom), p116, p127 (top), p132 (top).

Coo-ee Historical Picture Library: p6, p12, p14, p24, p26, p28 (top), p46 (bottom), p62, p64, p65 (bottom), p82, p88, p94, pp112-13, p131.

CSIRO Archives: p67, p126, p127 (bottom), p132 (bottom).

Dairy Farmers: p31.

NSW State Rail: p117.

Queensland Sugar: p42, pp48–9, p46 (top), p47, p48, p49.

Rural Press: p2, p15, p16, p17 (top), p18, p20 (top), p27, p35, p36, p37, p65 (top), p66 (bottom), p74 (bottom), p83, p84 (bottom), p85, p86, p90, p93, p96 (bottom), p97 (top), p102, p103, p104, p105, p106, p107, p108, p109, p110 (top), p114, p118, p119, p121, p122, p128, p129,

and for the photographs from The Land Heritage Collection used in Part Three at the foot of each company's or organisation's address panel.

Sydney Fish Market Limited: p75 (bottom), p76 (bottom).

Tyrrells Wines: p50, p56 (inset), p59.

Wildlight Photo Agency: cover, pp8–9, p10, p13, p80, p100–01, p134–35.

Woolworths Limited: p68 (bottom), p69.